World Heritage
Benefits Beyond Borders

Published on the 40th anniversary of the World Heritage Convention, this thematic collection of case studies provides a thorough understanding of World Heritage sites and their outstanding universal value in the context of sustainable development.

The case studies describe twenty-six thematically, typologically and regionally diverse World Heritage sites illustrating their benefits to local communities and ecosystems and sharing the lessons learned with the diverse range of stakeholders involved.

The volume emphasizes a holistic and integrated view of World Heritage, linking it to the role local communities play in management and protection, to issues of ecosystem sustainability, and the maintenance of biological, linguistic and cultural diversity.

Cross-disciplinary in its scope, this book will provide a meeting point for researchers, practitioners, community representatives and the wider public, and will promote cultural and natural heritage conservation as a key vector of sustainable development and social cohesion.

Direction and Concept
Kishore Rao, Director, UNESCO World Heritage Centre
Managing Editor
Vesna Vujicic-Lugassy, UNESCO World Heritage Centre
Photo Research and Coordination
Katerina Markelova, UNESCO World Heritage Centre
Photo Research Assistant
Clara Schoumann, UNESCO World Heritage Centre
Copy Editor
Caroline Lawrence

Volume Editor
Amareswar Galla

Acknowledgements
UNESCO expresses its deepest gratitude to the Government of Japan for its substantial financial contribution and generous cooperation provided in the production of this publication.

World Heritage
Benefits Beyond Borders

Edited by Amareswar Galla

United Nations
Educational, Scientific and
Cultural Organization
·
·
·
·
·
World
Heritage
Convention

United Nations
Educational, Scientific and
Cultural Organization
·
·
·
·
·
Supported by
Japanese Funds-in-Trust
to UNESCO

UNIVERSITY OF WINCHESTER
LIBRARY

CAMBRIDGE UNIVERSITY PRESS
Cambridge, New York, Melbourne, Madrid, Cape Town,
Singapore, São Paulo, Delhi, Mexico City

Published jointly by the United Nations Educational, Scientific and Cultural Organization
(UNESCO), 7, Place de Fontenoy, 75007 Paris, France, and Cambridge University Press,
The Edinburgh Building, Shaftesbury Road, Cambridge CB2 8RU, United Kingdom.

www.cambridge.org
Information on this title: www.cambridge.org/9781107610750

First published in 2012

A catalogue record for this publication is available from the British Library

ISBN UNESCO   978-92-3-104242-3 Paperback
ISBN Cambridge   978-1-107-61075-0 Paperback

Cambridge University Press has no responsibility for the persistence or
accuracy of URLs for external or third-party internet websites referred to
in this publication, and does not guarantee that any content on such
websites is, or will remain, accurate or appropriate.

The designations employed and the presentation of material throughout this publication
do not imply the expression of any opinion whatsoever on the part of UNESCO
concerning the legal status of any country, territory, city or area or of its authorities, or the
delimitation of its frontiers or boundaries.

The authors are responsible for the choice and the presentation of the facts contained
in this book and for the opinions expressed therein, which are not necessarily those of
UNESCO and do not commit the Organization.

Contents

v

Section 3   Integrated Planning and Indigenous Engagement

Section 4   Living Heritage and Safeguarding Outstanding Universal Value

Section 5   More than the Monumental

vii

Foreword

IRINA BOKOVA – DIRECTOR-GENERAL OF UNESCO

The 40th anniversary of the 1972 World Heritage Convention is an opportunity to take stock of achievements and to chart a new course for the future.

For four decades, the World Heritage Convention has helped to safeguard extraordinary places around the world for the enlightenment and enjoyment of present and future generations. In so doing, the Convention has drawn a new map of the globe. This map shows the bridges that link societies, blurring the geographical boundaries between countries and illustrating the intimate relationship between culture and nature. This is a map for peace, and a network for cultural exchanges that crosses the planet. To date, 190 States Parties have rallied around the simple but revolutionary idea that humanity is custodian to heritage of 'outstanding universal value' that must be protected for the benefit of all.

Many World Heritage sites carry iconic status – but it is time to look at them again, in order to forge new directions for their conservation and development. This volume seeks to explore sustainability as the key goal for heritage management, today and in the future. The concept of 'outstanding universal value' has meaning only if it is embedded in a local ecology, in harmony with local communities, with biological and cultural as well as linguistic diversity.

This volume unpacks this concept through twenty-six case studies that show the commitment by States, local authorities and communities to conserving and safeguarding their heritage. This is critical for the credibility of the World Heritage Convention and its future. Each study illustrates the way communities, site managers and other actors work to bring together people and their heritage using World Heritage to meet the needs of both conservation and development.

World Heritage carries local meaning, but its stakes are global. As we debate the contours of a new global sustainability agenda, we must ensure that culture has a central place, as both a driver of sustainable development and a source of inspiration and hope. Cultural heritage is the way we understand the world and the means by which we shape it. It is rooted in our cultural identities and provides a source of wisdom and knowledge to strengthen sustainable development policies and practices.

These are the horizons as we look to the future of World Heritage. Heritage is a motor for people's empowerment and sustainability – we must recognize this, understand it better and make the most of it, for local communities and humanity as whole.

Irina Bokova

Acronyms

AFD	Agence Française de Développement / French Development Agency
APNRM&L	Angkor Participatory Natural Resource Management & Livelihood (Cambodia)
APP	Área de Preservação Permanente / Permanent Preservation Area (Brazil)
APSARA	Authority for the Protection and Management of Angkor and the Region of Siem Reap (Cambodia)
CAP	Community Access Point (Australia)
CER	Collegium Educationis Revaliae (Estonia)
COMPACT	Community Management of Protected Areas Conservation Programme (UNDP)
CRAterre	Center for the Research and Application of Earth Architecture (France)
CRMD	Chief Roi Mata's Domain (Vanuatu)
CWSS	Common Wadden Sea Secretariat (Germany)
DANE	Departamento Administrativo Nacional de Estadística / National Administrative Department of Statistics (Colombia)
DED, currently GIZ	German Development Service
DFID	Department for International Development (UK)
DOLE	Department of Labor and Employment (Philippines)
EHESS	École des Hautes Études en Sciences Sociales (France)
ENAG	Lao National School of Administration and Management
EPA	Environmental Protection Agency (Yemen)
EU	European Union
FAO	Food and Agriculture Organization of the United Nations
FUMDHAM	Fundação Museu do Homem Americano / Museum of the American Man Foundation (Brazil)
GA	General Assembly (United Nations)
GBR	Great Barrier Reef (Australia)
GBRMPA	Great Barrier Reef Marine Park Authority
GBRWHA	Great Barrier Reef World Heritage Area
GDP	gross domestic product
GEF	Global Environment Facility
HABITAT	United Nations Human Settlements Programme

HRBA	Human Rights-Based Approach
HUL	Historic Urban Landscapes
IADB	Inter-American Development Bank
IBAMA	Brazilian Environment and Renewable Natural Resources Institute
ICOMOS	International Council on Monuments and Sites
IDP	Integrated Development Plan (South Africa)
IGO	intergovernmental organization
ILO	International Labour Organization
ILUA	Marine Park Indigenous Land Use Agreement (Australia)
IMACOF	Institut de Milieu Aquatic et Corridor Fluviale (France)
INDH	National Initiative for Human Development (Morocco)
INGO	international non-governmental organization
IUCN	International Union for Conservation of Nature
KEM	Khmer Effective Micro-organisms (Cambodia)
LAWHF	Local Authorities World Heritage Forum (UK)
LCB	Local Consultative Body (Senegal)
LEAP	Integrated Community Development and Cultural Heritage Site Preservation through Local Effort in Asia and the Pacific
LMAC	Local Marine Advisory Committee (Australia)
MDG-F	Millennium Development Goals Achievement Fund
MoA	Memorandum of Agreement
NBI	Necesidades Básicas Insatisfechas / Unsatisfied Basic Needs (Colombia)
NGO	non-governmental organization
NIKE	Nurturing Indigenous Knowledge Experts (Philippines)
OMVS	Organization for the Development of the Senegal River
PCD	Communal Development Plan (Morocco)
PCW	Pinelands Creative Workshop (Barbados)
PDP	Physical Development Plan (Barbados)
PEMP	Special Management and Protection Plan (Colombia)
POPs	persistent organic pollutants
PSMV	Plan de Sauvegarde et Mise en Valeur / Conservation and Enhancement Plan (Lao PDR)
QPWS	Queensland Parks and Wildlife Service (Australia)
RADEEMA	Autonomous Agency of Water Supply and Electricity of Marrakesh
RBTDS	Transboundary Biosphere Reserve of the Senegal River Delta
RTO	Rice Terraces Owners (Philippines)
SCOT	Scheme for Coherent Territorial Development (Lao PDR)
SDLIC	Sustainable Development in Low-Income Communities (Barbados)
SGP	Small Grants Programme (UNDP)

xii

SIDS	Small Island Developing States
SIPAR	Private Initiative Support for Aid to Reconstruction (Cambodia)
SITMo	Save the Ifugao Terraces Movement (Philippines)
SME	small and medium-sized enterprise
TEK	Traditional Ecological Knowledge
TUMRA	Traditional Use of Marine Resources Agreement (Australia)
UFPEl	Universidade Federal de Pernambuco (Brazil)
UN	United Nations
UNDG	United Nations Development Group
UNDP	United Nations Development Programme
UNDRIP	United Nations Declaration on the Rights of Indigenous Peoples
UNEP	United Nations Environment Programme
UNESCO	United Nations Educational, Scientific and Cultural Organization
UNF	United Nations Foundation
UNFPA	United Nations Population Fund
UNIDO	United Nations International Industrial Development Organization
UN-SCDP	Socotra Conservation and Development Programme (Yemen)
UNWTO	United Nations World Tourism Organization
UWI	University of the West Indies (Barbados)
WHTC	World Heritage and Tourism Committee
WCMC	World Conservation Monitoring Centre (UNEP)
WSP	Wadden Sea Plan

Introduction

VOLUME EDITOR

This volume has been published as a milestone, accessible to the wider public, on the occasion of the 40th anniversary of the World Heritage Convention. *World Heritage: Benefits Beyond Borders* is a thematic collection of case studies of World Heritage sites providing an understanding of their outstanding universal value in the context of sustainable development.

The publication is cross-disciplinary in scope, a meeting point for natural and social scientists, researchers and practitioners, professionals and community representatives. The twenty-six case studies represent a global spread of constructive and engaging examples. They have been selected on the principle of representativeness: outstanding universal value; inscription criteria; economic, social and environmental sustainability; inscriptions as natural, cultural and mixed sites; landscape as well as scientific and industrial heritage; and a regional balance of examples from around the world taking into consideration environmental, linguistic and cultural diversity.

Each case study assesses what is important for sustainable development with regard to the World Heritage site concerned; the management framework required for ensuring and enabling sustainable development and community engagement; benefits to local communities and ecosystems; lessons for sharing with other World Heritage sites; and the anticipated way forward in bringing together local and neighbouring communities through the environmental, economic and social dimensions of sustainability. As far as possible, evidence-based benefits are presented by the authors, who have written in the spirit of the call for transformations by the UNESCO Director-General: 'integrated cooperation mechanisms and more participatory governance structures for culture', 'deeper statistical understanding of the importance of the cultural sector to development' and 'greater awareness-raising about the cultural dimension of development'.[1]

The case studies are based on both published and unpublished material, as each is a critical reflection based on a synthesis of existing sources. The authors come from a range of culturally and linguistically diverse backgrounds. The first voice and idiom of each text has been ensured as well as possible, in the spirit of the participatory democracy that has become at once aspirational and quintessential in the implementation of the Convention. The value of the contributions is beyond the content, perspectives, methods and benefit-sharing illustrations. Readers are urged to further scope the possibilities for the safeguarding of the outstanding universal value of World Heritage sites in promoting sustainable economic, social and environmental development, cross-cultural understanding, and valuing heritage through both qualitative and quantitative indicators and seamless engagement to further benefits to communities beyond the site borders.

There is currently no publication of this type dealing with the issue of World Heritage and sustainable development through case studies. It will complement the existing literature on World Heritage which focuses on specific types of sites or specific issues, and will provide a broader, multi-issue context for understanding World Heritage. One of the strengths of the volume is its emphasis on a more holistic and integrated view, linking World Heritage to the role that local communities play in its management and protection, and to issues of ecosystem sustainability, management obstacles and possibilities, and the maintenance of biodiversity, as well as linguistic and cultural diversity.

The case studies have been grouped into five themes that address the concerns of safeguarding the outstanding universal value of World Heritage sites in the 21st century. One of the major challenges in the original drafting of the Convention, as well as in the current implementation, is to bring nature and culture under the same umbrella. Considerable progress has been made in bridging the nature / culture divide in heritage conservation. Illustrative case studies are presented under the title *Bridging Nature and Culture*.

Several World Heritage sites are concerned with the genesis of urban centres dealing with the history and development of particular complexes. They also deal with the process of urbanism covering the organic evolution and continuation of the urban centre itself. In the coming decade we will be crossing another major threshold in the history of humanity since the emergence of the first urban settlements in Mesopotamia and western Asia over five millennia ago. More than

[1] Points taken from speech by Irina Bokova at Diversity of Cultural Expression: Ministerial Forum of the Asia-Pacific Region, Dhaka (Bangladesh), 9 May 2012.

half the population of the world will be living in cities and towns. Globalization and the rapid growth of the world economy are accelerating the pace of urban development. In this context the safeguarding of the outstanding universal value of sites has come under severe pressure. Case studies presented under the title *Urbanism and Sustainable Heritage Development* illustrate a range of approaches to the conservation and sustainable development of World Heritage sites.

The United Nations Permanent Forum on Indigenous Issues, its Expert Mechanism on the Rights of Indigenous Peoples and the African Commission on Human and Peoples' Rights have advocated obligations under the United Nations Declaration on the Rights of Indigenous Peoples. The participatory process aimed at addressing the concerns and aspirations of indigenous peoples and stakeholder communities is illustrated by the group of texts entitled *Integrated Planning and Indigenous Engagement*.

The harmonization of soft law and hard law in the international field of standard setting for culture and heritage is crucial to ensure cooperation and coordination and economies of scale in the implementation by States Parties, INGOs and NGOs. In meaningful and sustainable community engagement at World Heritage sites the living heritage of stakeholder communities and their taking ownership is crucial for safeguarding the outstanding universal value of the sites, hence the case studies on *Living Heritage and Safeguarding Outstanding Universal Value*.

Finally, in the effort to promote a people-centred approach to conservation and to balance it with a site-centred approach, readers are urged to appreciate the benefits to local communities that are often not immediately visible, especially at large sites. Selected case studies are brought together under the title *More than the Monumental*.

The journey of four decades is without an end. While much has been achieved, with almost a thousand sites on the World Heritage List, the challenges are diverse and the achievements lead along multiple pathways. While inscription is a strategy with shared responsibility, conservation is an ongoing process. The various approaches to implementation of the World Heritage Convention, the most popular of the suite of UNESCO Conventions, have been strategic, innovative and inspirational. The commitment of States Parties to the Convention is commendable and the sense of ownership by local stakeholder communities has been heartening, as illustrated in the range of case studies compiled in this volume. As frequently emphasized, benefits from World Heritage status must accrue to local populations. The realization of this goal means transformations in heritage conservation that include local communities in social, economic and environmental sustainability.

1

Bridging Nature and Culture

The past four decades have witnessed a paradigm shift in the way the gulf between nature and culture has been bridged by managers and local communities in the conservation of World Heritage sites. In fact, the World Heritage Convention itself is the fundamental unifying framework for natural and cultural heritage conservation and this was further underscored by the World Heritage Committee, which adopted, in 2005, a unified set of World Heritage criteria following a first expert meeting on the subject in 1998. In according the respect due to global cultural diversity and different world views, the stakeholders have come to develop and practise a holistic ethic of conservation in bringing together people and their heritage across the binary of nature / culture divide. The range of case studies in this chapter illustrates this transformation. The local knowledge systems and communities practising heritage conservation on the ground have historically dealt with both nature and culture, often taking a systems approach, and they continue to do so.

Djoudj National Bird Sanctuary (Senegal), in particular, has witnessed the implementation of new mechanisms that put local communities and their integrated knowledge of nature and culture at the centre of government conservation priorities and concerns.

The overall economic value of the Great Barrier Reef (Australia) and its adjoining catchment area has been estimated to exceed AU$5.4 billion per annum and generates some 66,000 jobs, mostly in tourism. Over 220 Traditional Owners have undertaken compliance training, which has led to greater knowledge and awareness of marine compliance issues and, importantly, an increased feeling of empowerment by Traditional Owners managing sea country.

In Škocjan Caves (Slovenia), during a major annual festival, community members and cave managers present their conservation work and organize guided tours. The festival has become a joint activity of the park

Pond near Homhil village. Socotra is crucial to biodiversity conservation in the Horn of Africa.

management and the community to promote local production, encourage the use of local resources and revive traditional methods and customs.

In the Socotra Archipelago (Yemen), many plants are now being screened for medicinal properties, which have been used by local people for centuries, leading to a vast ethnobotanical knowledge that is deeply embedded in the local language.

In the Vega Archipelago (Norway), as early as the 9th century, tending eider ducks was reported to be one way that people made a living. The site was the core area for this tradition. Women played a key role, so World Heritage status also celebrates their contribution to down production.

1

Conservation of World Heritage and community engagement in a transboundary biosphere reserve: Djoudj National Bird Sanctuary, Senegal

TERENCE HAY-EDIE, KHATARY MBAYE AND MAMADOU SAMBA SOW[1]

Shared ecosystem

The close links between the lifestyles of local communities and natural resources have in recent decades led many policy-makers in Africa to review outdated management strategies adopted for the conservation and sustainable use of ecosystem services. In many countries, such as Senegal and Mauritania, the creation of protected areas in landscapes shared by local communities has been a source of conflict between the communities and government extension officers who hitherto relied on 'top-down' forms of management. Increasingly, stakeholder populations have viewed such actions as dispossession of livelihoods and disruption of access to the sustaining and socio-cultural services provided by the protected areas.

Senegal and Mauritania, in particular, have witnessed the implementation of new mechanisms that put local communities at the centre of government conservation priorities and concerns. In 1996, the introduction of national decentralization policies has allowed for the transfer of natural resource management to local communities. In the specific case of the Senegal River, this impetus resulted in the creation in 2005 of the Transboundary Biosphere Reserve of the Senegal River Delta (RBTDS).[2]

Senegal and Mauritania, in particular, have witnessed the implementation of new mechanisms that put local communities at the centre of government conservation priorities and concerns.

[1] Terence Hay-Edie, UNDP/GEF Small Grants Programme, programme advisor, New York; Khatary Mbaye, COMPACT local coordinator, Senegal; Mamadou Samba Sow, COMPACT local coordinator, Mauritania.

[2] Each protected area is a department entrusted with technical rules of management and access to resources by well-defined legal instruments (i.e. forest code, environmental code, hunting code, fishing code). In the case of RBTDS, governance bodies including national and transnational committees are set up but the local communities are still marginal in decision-making.

7

COMPACT stakeholders,
meeting in February 2011.

Dating from the colonial period when Saint-Louis, former capital of French West Africa (1895–1958), was based at the mouth of the Senegal River, a range of hydro-agricultural infrastructure projects was developed with a view to improving agricultural production (in particular for the intensive planting of peanuts and other cash crops), and moderating the effects of successive droughts in the region. Most recently, in 1986, the Djama dam was built to control and moderate the hydrological regime of the Senegal River, as well as to increase the availability of fresh water needed for irrigated agriculture. Prior to the introduction of the upstream dam, river flow depended entirely on rainfall and flooding cycles, and could not be regulated.

Under the natural hydrological regime, salt-water intrusion previously extended as far as 200 km inland from the mouth of the Senegal River, during the dry season; contrasted with high levels of flash flooding during the rainy season (negatively affecting the Island of Saint-Louis, also listed under the World Heritage Convention for its cultural values). However, while beneficial in many respects, the Djama dam also resulted in unanticipated ecological impacts on the Djoudj/Djawling delta ecosystem, most notably through the proliferation of invasive aquatic weeds which today constitute one of the main threats to the habitats of wildlife and the outstanding universal value of the World Heritage site.

Dry conditions in park headquarters. The construction of the Djama dam threatens the annual wet–dry cycle that brings life to the Djoudj Sanctuary.

Natural and cultural values of the transboundary ecosystem

Following a lengthy political and institutional process involving actors ranging from the governments of both countries, representatives of civil society, and development partners, the RBTDS was created in 2005 with a total area of 641,768 ha.[3] The RBTDS is characterized by a diversity of ecosystems along with a complex of wetland areas which together host significant birdlife. The transboundary ecosystem within the RBTDS encompasses Djoudj National Bird Sanctuary (the core zone of the UNESCO World Heritage site, and a wetland of international significance under the Ramsar Convention), Réserve de Ndiaël, Réserve de Gueumbeul, Langue de Barbarie and the wildlife reserve of Chatt Boul (on the Senegalese side); and Djawling National Park (on the Mauritanian side). The shared ecosystem between the two countries is separated by the Senegal River and connected by a network of artificial dykes and ecological corridors to allow for the migration of fauna from one side to the other.

The delta is known particularly for the presence of an estimated 1.5 million birds including a number of migratory birds (including large flocks of flamingos and pelicans) which spend the winter in Africa. Terrestrial fauna is composed of reptiles (turtles, snakes, lizards) and mammals (warthogs, jackal, monkeys). The range of fish includes both freshwater and estuarine species. Several species of amphibians and crustaceans are also observed. The protected areas of Saint-Louis Marine Park and the Langue de Barbarie are also home to important nesting sites for marine turtles (i.e. leatherback turtle, *Dermochelys coriacea*; green turtle, *Chelonia mydas*; and loggerhead turtle, *Caretta caretta*).

Flora contains stands of *Acacia nilotica, Prosopis* and *Tamarix* as well as several economically useful fibre species such as *Sporobolus robustus*, used by women for the production of local fine hand-woven mats (which are sought after across the region). The local craft industry is directly linked to the development of natural resources using customary practices and traditional ecological

---

[3] The RBTDS includes 562,470 ha of terrestrial and 79,298 ha of maritime areas. This covers an area of 186,908 ha on the Mauritanian side and 454,860 ha on the Senegalese side.

The shared ecosystem between Senegal and Mauritania is divided by the Senegal River and connected by a network of artificial dykes and ecological corridors to allow for the migration of fauna from one side to the other.

10

knowledge (TEK) derived from artisanal processing, including leather goods such as bags, shoes and souvenirs.

The main current threats to biodiversity, following the commissioning of the Djama dam, have resulted from the permanent availability of fresh water (i.e. which is no longer marked by natural annual cycles of intermittent fresh and salt-water flooding) leading to the proliferation of invasive aquatic plants, such as *Typha australis*, *Pistia stratiotes* and *Salvinia molesta*. The human management of water and modification of drainage flows into different channels have disrupted and greatly reduced bird habitats. Some areas of the Djoudj and Djawling have become clogged by dense invasive weeds, and cannot be accessed in order to achieve effective monitoring and viewing of wildlife.

The new hydrological facilities have also considerably reduced traditional areas of pasture, resulting in an increase of incursions by livestock into the Djoudj and Djawling parks' core zones. Although access to these areas is prohibited, poverty and a lack of alternative sources of income have also led local populations to frequently practice illegal fishing, and the illegal collection and use of firewood. In addition, inadequate investment in the park infrastructure has also been one of the reasons for technical inefficiency in managing the national park.

The city of Saint-Louis continues to provide important opportunities for tourism, with a vibrant hotel sector. However, despite the opportunities for camping and ecotourism the local communities have yet to take full advantage of this sector in order to diversify their economic base. The city of Keur Macène, which is the largest on the Mauritanian side, has more limited infrastructure, so local populations depend mainly on Saint-Louis as a regional market. Community identity in the villages on both sides of the river is composed of common ethnicity composed of Wolof, Moors and Fulani, who share matrimonial and

Women craftworkers using native plants to make high-quality mats for sale (Mauritania).

commercial relations, presenting a significant advantage in promoting cross-border cooperation, integration of project development and governance for the RBTDS.

Transboundary cooperation between Senegal and Mauritania

The Organization for the Development of the Senegal River (OMVS) is the main agency responsible for the administration and regulation of the hydrological system under the subregional cooperation framework between Mauritania and Senegal. Both Djawling National Park and Djoudj National Bird Sanctuary depend on the OMVS for the regulation of the river ecosystem under the legal instrument of the Water Management Charter. In addition, the West African Regional Marine and Coastal Conservation Programme[4] has incorporated elements of both Djawling National Park, in Mauritania, and the Langue de Barbarie and marine protected area of Saint-Louis, in Senegal.

In this regard, with financing from the Global Environment Facility (GEF), UNEP and UNDP, a project on the conservation of biological diversity through the rehabilitation of arid and semi-arid areas along the border between the two countries occurred between 2001 to 2005, further helping to build cooperation between the two countries in policy management, integration and harmonization of interventions in the Senegal River basin. However, ongoing degradation of the water management infrastructure (i.e. dykes and waterways) necessary to comply with the watershed management plans in the RBTDS, continues to pose significant challenges with regard to the full implementation of the established intergovernmental cooperation framework.

Participation of local communities in the RBTDS

The involvement of local communities in managing the natural resources has significantly evolved in West Africa in recent decades. In the case of the

[4] http://en.prcmarine.org/

11

RBTDS, key opportunities have been seized to initiate cross-border coop- eration activities between communities giving birth to 'twinning' arrange- ments between the two parks. Local communities have played a crucial role in the process of creating the RBTDS, as well as in the improvement of the visibility of the outstanding universal value of the area as a globally signifi- cant protected landscape.

Between 2000 and 2004, the GEF Small Grants Programme (GEF SGP),[5] implemented by the UNDP in Senegal, began supporting a number of local initiatives in the RBTDS to address threats to Djoudj National Bird Sanctuary, in particular through the fight against *Salvinia molesta*. Beginning in 2000, SGP Senegal supported the Association Diapanté, a local community-based organization who partnered with an NGO, the Civil-Military Committee to Support Development, to organize teams of volunteers to go out on the river and pull out the invasive plants *Salvinia molesta* to protect the waterways and channels for birdwatching and other community-based ecotourism activities.

During the project, local village chiefs signed agreements with the project management committee to support the clean-up efforts. Diapanté has also established a small revolving fund to finance micro-projects led by local communities. Through the revolving fund, the women of Médina Maka (a village near Djoudj National Bird Sanctuary), who had been strong participants in efforts to remove the invasive species, were supported in establishing liveli- hood activities, such as agroforestry and producing incense. The efforts yielded impressive results: over 27,000 m² were cleaned in more than ninety sites along the river, representing the product of 6,262 'people-days' of work. Diapanté also received numerous additional requests for micro-project lending from the revolving fund in order to keep the conservation activities running.

In 2007, with financial support from the GEF and the United Nations Foundation, the SGP pilot efforts in Djoudj National Bird Sanctuary were up-scaled when the Community Management of Protected Areas Conservation Programme (COMPACT) was initiated to support local community initiatives to conserve and promote biodiversity in the RBTDS.[6] As part of a compre- hensive landscape-level planning process developed for six World Heritage

12

Teams of volunteers were organized to go out on the river and pull out the invasive plants *Salvinia molesta* to protect the waterways and channels for bird- watching and other community-based ecotourism activities.

[5] The GEF SGP aims to deliver global environmental benefits in the GEF Focal Areas of biodiversity conservation, climate change mitigation, protection of international waters, prevention of land degradation (primarily desertification and deforestation), and elimination of persistent organic pollutants (POPs) through community-based approaches. http://sgp.undp.org/

[6] COMPACT aims to replicate the success of the GEF Small Grants Programme at the national scale for protected landscapes, including natural World Heritage sites and overlapping Biosphere Reserves recognized for their global outstanding universal value. http://sgp. undp.org/index.php?option=com_content&view=article&id=103:compact&catid=45:about- us&Itemid=165.

The vegetation of the Djoudj sanctuary reflects low rainfall, and halophytic plants cover much of the area.

sites elsewhere (Belize Barrier Reef Reserve System, Belize; Morne Trois Pitons National Park, Dominica; Mount Kenya National Park, Kenya; Sian Ka'an Biosphere Reserve, Mexico; Puerto Princesa Subterranean River National Park, Philippines; Mount Kilimanjaro National Park, Tanzania), a baseline analysis, conceptual model of key threats, and site strategy for COMPACT were developed involving the full range of RBTDS stakeholders. The stated aim of COMPACT in the RBTDS was to establish a range of partnerships and practical actions through a network of civil society-led conservation efforts.

As of early 2012, a total of twenty-two COMPACT small grants projects together worth over US$400,000 have been funded (with nineteen in Senegal and three in Mauritania) addressing a range of key threats and pressures identified through the conceptual modelling approach. By working through a network of interlinked local initiatives, a number of results have been achieved through the provision of carefully selected small grants, including for the rehabilitation of wildlife habitat; reducing pressure on natural resources; ecological monitoring; and promoting capacity-building.

As identified in the COMPACT conceptual model and site strategy, the rehabilitation of wildlife habitat constitutes a priority to reduce the threats on the outstanding universal value of the World Heritage site. In response, the Local Consultative Body (LCB), composed of a cross-section of local stakeholders responsible for the strategic direction of COMPACT at the level of the protected area, approved a range of small grants averaging US$25,000 (funded by GEF and UNF) to address invasive species and biodiversity protection. The SGP grants have *inter alia* addressed the following key areas of intervention:

■ 20 km of waterways, previously invaded by *Typha australis*, have been rehabilitated as bird habitat by the neighbouring village populations living in the vicinity of Djoudj National Bird Sanctuary in Senegal.
■ Building on the traditional knowledge of fishing communities along the coast in Djawling National Park in Mauritania, the passage of fish at the Berbar watergate has been facilitated to allow for greater ecological

13

connectivity, allowing juvenile fish to move between spawning pools across the shared Djoudj/Djawling ecosystem.

- A large pool in Ndiaël Special Wildlife Reserve was restored thanks to a 're-flooding' initiative conducted by the outlying villages. Numerous bird species which had not been seen in the area have now returned to the wetland.
- In Mauritania, a community-run nursery of 25,000 seedlings was organized to serve the rehabilitation of 6 ha of *Sporobolus robustus* fields, to be used to provide a sustainable supply of plant fibre for the women's weaving cooperatives marketing traditional floor-mats.
- In Senegal, an anti-erosive dam was built by community members to protect an endangered nesting and reproduction site for marine birds on an islet in the National Park of the Langue de Barbarie.
- Nesting sites for endangered marine turtles in the National Park of the Langue de Barbarie have been documented, mapped and protected by local community volunteers with a view to further developing community-based ecotourism activities.

In order to reduce pressure on natural resources in the Djoudj and Djawling parks, and wider landscape within the RBTDS, alternative sources of energy and biomass-based production have also been encouraged. In particular, forage species have been promoted along with improved agricultural, agroforestry and sustainable fisheries practices.

The SGP grants have *inter alia* addressed the following key areas of community livelihood and empowerment needs:

- 24 ha of fodder crops have been planted and sustained to fill the deficit and reduce the frequent incursions of livestock into Djoudj National Bird Sanctuary.
- Nine deposits of butane gas have been put in place with the assistance of the park authorities in the vicinity of the park, allowing over 400 households to access alternative energy sources to firewood.
- A framework of regular community consultations has been put in place to address cases of illegal fishing practices.
- In Mauritania, 150 women have been trained in sustainable use techniques and environmental plans, and 50 women in project management and environmental education.
- A revolving business development fund of US$56,000 (Association Française d'Ingénierie Système) has been established to support business enterprises to improve income-generating activities with the interest

Five thousand white pelicans nest regularly in the park.

generated. Concessions to date have been made for fifteen solar kits for the electrification of three villages in Ghahra, Sbeikhat and Bariel.

With regard to ecological monitoring, COMPACT has involved the local populations through the training of 'ecoguards'. Key elements of the park infrastructure in the Djoudj National Bird Sanctuary have been upgraded with the help of local communities including the command post, watchtowers, observation stations and panels for walking tracks. There are 160 CBO members of the villages who have tracking devices and who have had appropriate training to use them in the Djoudj and the World Heritage site.

With regard to ecological monitoring, COMPACT has involved the local populations through the training of 'ecoguards'.

Looking to the future from both banks of the river

In conclusion, the political willingness of the two States Parties to the World Heritage Convention to cooperate in the governance, integration, and management of the RBTDS across the Senegal River basin has been a leading example of transboundary conservation. The RBTDS cooperation framework also presents significant opportunities to build upon the shared environmental challenges on both banks of the river (i.e. invasive plants, hydrological connectivity, salinity, loss of biodiversity), and develop a range of specific on-the-ground implementation and zoning activities to ensure that the collaborative

A range of specific zoning activities was implemented to ensure that the collaborative framework between Mauritania and Senegal is put into practice.

framework is actually put into practice (Borrini-Feyerabend and Hamerlynck, 2011).

Some of the most remarkable results in the RBTDS in recent years have been achieved through the range of partnerships developed under the COMPACT programme since 2007. In particular, teams of 'ecoguards', consisting of both male and female volunteers chosen among local communities, first formed and supported by SGP and COMPACT, continue to ensure the ecological monitoring of the protected area in collaboration with the protected area authorities. In particular, the partnership between the government authorities and the local communities has resulted in the participation of village organizations in the annual bird survey (conducted in January each year).

In the case of Saint-Louis Marine Park and Ndiaël Special Wildlife Reserve, community members from outlying villages have also established 'wildlife management bodies' to work in collaboration with the government technical staff. Coupled with their involvement in the collection and provision of data, local populations who share cultural ties are also assisting the elaboration of management plans for different protected areas with the RBTDS through the development and understanding of 'participatory conceptual models' which graphically capture the ecosystem connectivities in the river delta.

Looking to the future, a common history uniting the Senegal delta populations by Saint-Louis (also recognized as a UNESCO World Heritage city), may act as a 'regional hub' for renewed socio-economic development, and continue

The sanctuary is a wetland of 16,000 ha, comprising a large lake surrounded by streams, ponds and backwaters.

to provide an exchange platform between the diverse ethnic communities of the area. As exemplified though the COMPACT approach, considerable scope exists to build on the transboundary cooperation framework by working directly with the related communities of Wolof, Fulani and Moorish who share strong trade relations and cultural affinities.

Since its inception in 2007, COMPACT has contributed to the conservation of the RBTDS by directly supporting community initiatives carried by the local populations through participatory methodologies. The cumulative impact of the network of inter-linked small grants has promoted a greater sense of civic ownership for the World Heritage site (including a better understanding of the outstanding universal value) by the local communities. Moreover, as shown through the *Journées du Delta* organized in January 2012, the results of the civil society-led conservation efforts are much in evidence as part of the ongoing review by the respective governments to put in place a new system to implement a shared co-management and governance arrangement within the RBTDS.

2

Community engagement in safeguarding the world's largest reef: Great Barrier Reef, Australia

JON C. DAY[1], LIZ WREN[1] AND KAREN VOHLAND[1]

Managing an icon

The Great Barrier Reef (GBR) along the north-east coast of Australia is the world's most extensive coral reef ecosystem and includes around 10 per cent of all the world's coral reefs. As one of the richest and most complex ecosystems on Earth, it is a significant global resource.

The GBR's outstanding universal value was recognized in 1981 when an area of 348,000 km[2] was inscribed on the World Heritage List,[2] meeting all four 'natural' criteria which at that time[3] were:

(i)   outstanding examples representing the major stages of the Earth's evolutionary history;

(ii) outstanding examples representing significant ongoing geological processes, biological evolution and <u>man's interaction with his natural environment</u>;[4]

(iii) unique, rare or superlative natural phenomena, formations or features or areas of exceptional natural beauty, such as superlative examples of the most important ecosystems to man;

(iv) habitats where populations of rare or endangered species of plants and animals still survive.

---

[1] Great Barrier Reef Marine Park Authority, Townsville, Australia.
[2] Equivalent to the size of Italy or Japan, or if placed on the west coast of the US, would stretch from the Canadian border to the Mexican border.
[3] Note that the wording and the numbering of the 'natural' World Heritage criteria have changed since 1981 so differ from the words in the *Operational Guidelines* today.
[4] The wording underlined above was an important inclusion in 1981 but no longer forms part of the natural World Heritage criteria and has since evolved into the concept of a 'cultural landscape'.

Marine Park Authority Board Member Melissa George congratulates Gurang Traditional Owner Kerry Blackman on the signing of the historic Port Curtis Coral Coast Traditional Use of Marine Resources Agreement (TUMRA), one of five operating in Queensland.

Today the Great Barrier Reef World Heritage Area (GBRWHA) still exhibits its remarkable biodiversity, including shallow inshore fringing reefs and mangroves, continental islands, coral cays, seagrass beds, mid-shelf reefs and exposed outer reefs, and extends out to deep oceanic waters more than 250 km offshore.

Within the boundary of the GBRWHA are:

- the GBR Marine Park – the Federal Marine Park comprises just under 99 per cent of the World Heritage Area. The GBR Marine Park's jurisdiction ends at low water mark along the mainland coast (with the exception of port areas) and around islands (with the exception of seventy Commonwealth-owned islands which are part of the Marine Park);
- some 950 islands within the jurisdiction of Queensland, about half of which are declared as 'national parks'; and
- internal waters of Queensland (including a number of significant port areas).

While no longer the largest World Heritage property in the world, the GBR remains arguably the most significant for biodiversity conservation. Coral reefs, mangroves and seagrass habitats occur elsewhere on the planet, but no other property covers such latitudinal and cross-shelf diversity (combined with diversity through the depths of the water column), which collectively means a globally unique array of ecological communities, habitats and species.

The management of such a large and iconic World Heritage property is complex due to the overlapping federal and state (Queensland) jurisdictions. Management therefore relies upon a number of federal and state government agencies working within a framework of an Intergovernmental Agreement (revised in 2009), using a combination of management tools (such as Zoning Plans, Plans of Management, Fishery Management Plans, Traditional Use of Marine Resources Agreements, Dugong Protection Areas, permits), along with various management approaches (including education, planning, environmental impact assessment, monitoring, stewardship and enforcement) to regulate access, and to control and/ or mitigate impacts associated with activities (such as tourism, fisheries, shipping) or to address pressures (such as climate change or declining water quality).

The various agencies involved in GBR management include:

- The Great Barrier Reef Marine Park Authority (GBRMPA) – the primary federal agency responsible for planning and management of the GBR

19

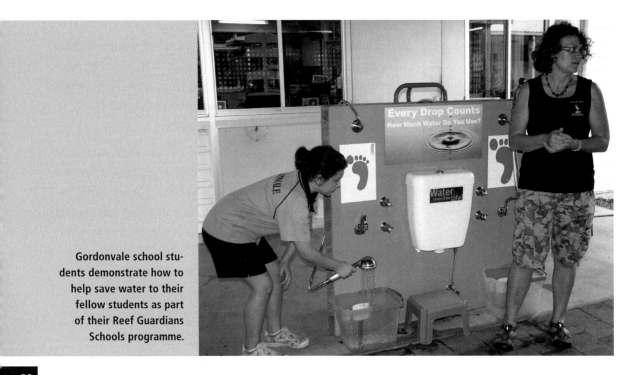

Gordonvale school students demonstrate how to help save water to their fellow students as part of their Reef Guardians Schools programme.

Marine Park. The GBRMPA is a statutory authority with its own federal legislation, and is responsible to the Federal Environment Minister.

- The Australian Government's Department of Sustainability, Environment, Water, Population and Communities – responsible for the regulation of activities that may have a significant impact on outstanding universal value, and other matters under the national environmental legislation; acts as the 'State Party' responsible for all Australian World Heritage Areas.
- Various Queensland agencies assist in the management of the GBR and the adjoining lands, islands and tidal waters, with the Queensland Parks and Wildlife Service (QPWS) having a major responsibility for field management activities.
- Other Australian and state government agencies are involved in specific aspects of management such as shipping, fisheries, defence training and aerial surveillance.

All the above agencies recognize that the GBRWHA is extremely important to indigenous and local communities as well as commercial and recreational users that depend on the GBR for recreation and their livelihoods.

The iconic status of the GBR leads many people to think the entire area is a marine sanctuary or a marine national park. However since declaration in 1975, the GBR has always been a multiple-use, marine-protected area in which

Lama Lama Indigenous Rangers learning to use an electronic 'i-tracker' to help map important sites within their sea country.

21

The economic and financial value of tourism, recreational activities, and commercial fishing within the GBR Marine Park (plus tourism in the adjoining catchment area) has been estimated to exceed AU$5.4 billion per annum and generates some 66,000 jobs, mostly in tourism.

zoning provides one of the key management tools. The multiple-use zoning approach provides for the separation of conflicting uses while allowing a wide range of commercial and recreational activities, some of which are further managed through a permit system.

The GBR's social and economic value is significant, particularly for the 1.12 million residents in the adjoining areas.

The economic and financial value of tourism, recreational activities and commercial fishing within the GBR Marine Park (plus tourism in the adjoining catchment area) has been estimated to exceed AU$5.4 billion per annum and generates some 66,000 jobs, mostly in tourism.[5]

In recent years, there have been an estimated 14.6 million recreational visits per year to the GBR from residents in the GBR catchment, with an additional 1.9 million visitor days from tourists carried by commercial operators into the GBR. Over 83,000 recreational vessels are registered in the GBR coastal areas.

These industries underpin a significant and growing proportion of Queensland's regional economy and they rely on the continued health of the

[5] A 2008 report by Access Economics provided quantitative estimates of the economic and financial value of tourism, commercial fishing and recreational activities undertaken in the Great Barrier Reef Marine Park and its catchment area. The majority of the GBR's contribution to the Australian economy of over AU$5.4 billion per annum comes from tourism.

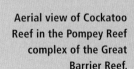

Aerial view of Cockatoo
Reef in the Pompey Reef
complex of the Great
Barrier Reef.

GBR ecosystem for long-term economic sustainability. Considerable efforts have been invested in establishing effective partnerships with the key industries; in particular the tourism industry and, increasingly, parts of the commercial fishing industry are recognizing that there are benefits in working closely with the managing agencies.

The GBR is, however, important for a broad range of local communities and stakeholders, not just those who use its resources. Consequently, there are many ways that community members work closely with the management agencies to ensure effective protection and management of the GBRWHA.

Fundamental role of local communities in the GBR

Since the early days of the GBR Marine Park in the late 1970s, there has been recognition of the importance of local communities to help protect the natural and cultural values. Today there is a very strong commitment to maintain effective and meaningful partnerships with indigenous people, local communities and users in order to conserve the outstanding universal value as well as enhance the resilience of the GBR.

Much of this community engagement has been ongoing since the declaration of the GBR Marine Park, but one example where additional resources

An elder of the Thiithar clan of the Guugu Yimmithirr Nation of Traditional Owners teaches children about their Sea Country at Elim Beach. The Thiithar, who live in the Hopevale region, became the first in Queensland to receive community land and the right to self-government in 1986.

were channelled into intense periods of public engagement in order to achieve important and specific outcomes was the participatory activities during the GBR-wide rezoning in the late 1990s.

Aboriginal and Torres Strait Islander people

For thousands of years Aboriginal and Torres Strait Islanders have had a traditional connection with the marine environment of the GBR region[6]. Traditional customs and spiritual lore continue to be practised in the sustainable use of sea country today. There are at least seventy Traditional Owner groups, from the eastern Torres Strait Islands to just north of Bundaberg, whose sea country includes parts of the GBRWHA. Each of these groups sustains a range of past, present and future cultural and heritage values for land and sea country and for surrounding sea countries.

Given the long association of Aboriginal and Torres Strait Islander people in the GBR Region, it is appropriate that World Heritage criterion (ii) in 1981 enabled reference to *'man's interaction with his natural environment'*.

---

[6] Evidence exists indicating the earliest occupation by indigenous people in northern Australia is somewhere between 40,000 and 60,000 years ago.

The GBRMPA has established an Indigenous Partnerships Group that works closely with Traditional Owners, acknowledging their continuing traditional connections to the GBR. This includes collaborating with Traditional Owner groups and with the Queensland government to develop a suite of sea country management arrangements including Traditional Use of Marine Resources Agreements (TUMRAs) and Marine Park Indigenous Land Use Agreements (ILUAs). TUMRAs are formal agreements developed by Traditional Owner groups and accredited by both the GBRMPA and Queensland.

A TUMRA may describe, for example, Traditional Owners' sea country aspirations, how Traditional Owner groups wish to manage their take of natural resources (including protected species), their role in compliance and their role in monitoring the condition of plants and animals, and human activities in their sea country. An implementation plan associated with a TUMRA may describe ways to educate the public about traditional connections to sea country, and to educate other members of a Traditional Owner group about the conditions of the TUMRA.

Over 220 Traditional Owners have undertaken compliance training, which has led to an increased knowledge and awareness of marine compliance issues, and importantly, an increased feeling of empowerment by Traditional Owners managing sea country.

There are currently five TUMRAs and one ILUA accredited in the GBR Marine Park which collectively cover some 18 per cent of sea country within the Marine Park (or 22 per cent of the coastline), and involve fourteen Traditional Owner groups. Each TUMRA operates for a set time after which it is renegotiated.

The Australian Government's Reef Rescue Program has provided an opportunity to enhance a range of sea country partnership activities including the sustainable traditional use of marine resources, indigenous tourism, sea country research and education, sea country planning and Marine Park compliance matters. It is projected that over the life of the sea country partnerships investment (to 30 June 2013), over fifty of the estimated seventy Traditional Owner groups connected to the GBR will be actively engaged in sea country management activities.

The programme also sees Traditional Owners working with the GBRMPA officers to marry their cultural knowledge with western knowledge to better protect the GBR. Over 220 Traditional Owners have undertaken compliance training, which has led to an increased knowledge and awareness of marine compliance issues, and importantly, an increased feeling of empowerment by Traditional Owners managing sea country.

The GBRMPA also fosters indigenous community engagement through membership on the Authority Board (one indigenous member is required by the legislation). An Indigenous Reef Advisory Committee has also been established to provide advice to the agency on how to best engage indigenous communities and Traditional Owners.

The Great Barrier Reef is one of the most significant World Heritage properties for biodiversity conservation. Picture shows a burrowing clam, sea squirts and hard coral.

Another successful example has been the Story Place database that shares information and knowledge about Traditional Owners and their relationship with land and sea country. It is a useful resource for Aboriginal and Torres Strait Islanders, managers, researchers and others interested in learning more about Traditional Owner connections with the GBR.

Local Marine Advisory Committees

Advice to the GBRMPA on management issues at a local level occurs through voluntary community-based committees called Local Marine Advisory Committees (LMACs). Established in 1999, the twelve LMACs operating along the GBR coast help the GBRMPA and other management agencies to keep in touch with marine and coastal issues at a local level and understand the use of the GBRWHA.

Members may be independent, or represent a community or industry group from which they coordinate feedback. The aim is to have a balanced representation of local people who are involved in the management or use of the Marine Park. Major benefits of LMACs include an opportunity for a two-way flow of information between the local community and the management agencies.

Composition of each LMAC varies depending on local interests and industries, and the 210 community members currently involved in LMACs along the GBR Coast bring perspectives from commercial and recreational fishers, conservation groups, farmers, tourist operators, local government, industry and interest groups as well as Aboriginal and Torres Strait Islander interests. Relevant federal and state government agency representatives, including one of GBRMPA's senior management team, also attend LMAC meetings.

Another successful way to engage local communities occurs through Community Access Points (CAPs); in excess of 300 CAPs along the GBR coast today provide zoning maps and information developed by the management agencies direct to users and to interested stakeholders.

Reef Guardian Program

The GBRMPA began its Reef Guardian concept in 2003 with Reef Guardian Schools. Today the Reef Guardian Program proves that a hands-on community level approach, through building relationships and working with those

25

who use or rely on the GBR for their businesses or for recreation, can help to protect the GBR's social, economic and environmental values and build a healthier and more resilient GBR.

The Reef Guardian Schools Program continues to go from strength to strength with over 111,000 students from 285 schools including state, Catholic and independent schools across Queensland now involved in stewardship programmes which help to look after the GBR. Given that the entire population of the catchment adjoining the GBR is around 1.12 million people, it is significant that 10 per cent of the entire population adjoining the GBR are now involved in this programme.

Given the success of the Reef Guardian Schools Program, the initiative was expanded to include Local Government Councils in 2007. Currently, thirteen councils from Bundaberg to Cooktown are signed up as Reef Guardian Councils, effectively demonstrating the actions they are taking to improve the health and resilience of the GBR through initiatives such as water management, urban storm water treatment, waste reuse and recycling, vegetation and pest management, erosion control, land use planning, energy and resource efficiency, and community education.

In 2010 some additional short-term funding was provided to strengthen the Reef Guardian Councils and Schools programmes as well as expand the Reef Guardian initiative to include pilot programmes for Reef Guardian Farmers and Reef Guardian Fishers. Although still in the early stages of development, the Fishers and Farmers programmes help to promote the valuable initiatives being undertaken by individuals across these industry sectors – over and above what is required by regulations and legislation – which are delivering significant environmental benefits.

All these initiatives aim to foster the sharing of information and the uptake of actions that will help to improve the social and economic sustainability of these users and industries, and ensure the environmental sustainability of the GBRWHA by improving its health and resilience.

Role of the community in rezoning

One example of specific targeted community engagement was part of the comprehensive planning programme that resulted in the rezoning of the entire marine park from 1999 to 2003. This included one of the most comprehensive processes of community involvement and participatory planning for any environmental concern in Australia's history. The consultation included two formal phases of public input and submissions, and ongoing interaction

The Great Barrier Reef contains six of the world's seven species of marine turtles, and includes the largest green turtle nesting area in the world. Picture shows a loggerhead turtle.

with key stakeholders and communities throughout the planning process.

The public consultation programme included some 1,000 formal and informal meetings as well as information sessions with people along and beyond the GBR coast. These meetings included local communities, commercial and recreational fishing organizations, indigenous people (including many Traditional Owners), tourism operators, conservation groups, local councils, and state and federal politicians.

A wide range of products were developed to engage the various community groups, aimed particularly to communicate the Draft Zoning Plan to users and to obtain constructive feedback. As a result, 31,600 written public submissions were received during the entire rezoning process, which led to considerable changes in the final outcome without compromising the primary aim of better protecting the biodiversity of the GBR.

27

Safeguarding the outstanding universal value

While the GBR remains one of the healthiest coral reef ecosystems on the planet, its condition has declined significantly since European settlement. The *Great Barrier Reef Outlook Report 2009* concluded that the GBR ecosystem is at a crossroad, and decisions made in the next few years are likely to determine its long-term future.

The *Outlook Report* identified the four priority issues facing the GBR as: climate change; continuing decline in water quality from catchment runoff; loss of coastal habitats from coastal development; and a small number of impacts from fishing activities including the incidental catch of protected species and the death of non-targeted species.

The best way to ensure the future of the GBR is to reduce as many of these pressures as possible, allowing the ecosystem to be more resilient to the pressures that remain. Building and maintaining resilience is central to protecting the GBRWHA, and in some of these areas major advances have occurred. For example, through the Reef Water Quality Protection Plan, the Australian and Queensland Governments are taking significant steps and contributing substantial investment to improve the quality of water flowing from agricultural lands into the GBR coastal waters.

Emerging issues that have become apparent since the 2009 *Outlook Report* will be considered as part of the assessment process for the next (2014) report. These include: increases in shipping activity as a result of port expansions; population growth as a result of urban and industrial activities along the coast; coastal development and changes in land use within the GBR catchment; and extreme weather events including floods and cyclones.

With increasing population growth and an expanding mining industry in GBR catchments, it is anticipated that coastal development will remain a priority management issue requiring greater focus in coming years. As a result, the comprehensive strategic assessments currently under way and other continuing improvements to management of the coastal zone will be critical in improving the outlook for the GBR.

Lessons learned/challenges

Getting local communities involved in both protection and management has been one of the success stories in the GBR. The Zoning Plan which came into effect in mid 2004 included many changes between the draft and final plans resulting from the huge public input.

There are benefits to both the GBRWHA and to local communities being inscribed on the World Heritage List. World Heritage status promotes local and national pride in the GBR, which translates into feelings of responsibility to protect the area. However listing also means greater scrutiny and accountability, given the internationally acknowledged importance of the area.

The specific political, economic, social and managerial context of the GBRWHA must be taken into account when considering whether any of the lessons from the GBR might be translated to other World Heritage areas. Furthermore, the ongoing management of the GBR is effective for a range of reasons, some of which may not be relevant or achievable in other areas. Day (2011) lists some key aspects of the GBR management context, including widespread consensus that the GBR is important, with many industries and stakeholders recognizing it is worth conserving. This then translates to consequent socio-political support.

Other factors that contribute to the success of the current adaptive management and ecosystem-based approach include a relatively low population along the GBR coast, a reasonably high standard of living, and well-established and relatively stable governance at all levels of government. Indigenous and local communities also play a fundamental role in ensuring sustainable development.

28

World Heritage status promotes local and national pride in the GBR, which translates into feelings of responsibility to protect the area. However listing also means greater scrutiny and accountability, given the internationally acknowledged importance of the area.

The Great Barrier Reef comprises the world's most extensive coral reef ecosystem and includes around 10 per cent of all the world's coral reefs. Picture shows a large soft coral.

However there is no doubt that an informed and involved community is essential if the GBRWHA is to be used and managed in a way that ensures that Australia meets its obligations under the World Heritage Convention. For protection of the outstanding universal value to remain successful, it must be undertaken within a collaborative framework that recognizes the close relationship between sustainable community livelihoods and the effective protection and management of the GBR.

Successful engagement is dependent on the willingness of community members and stakeholders to engage on matters that are important to them, and on the level of commitment of managers to also get it right. In doing so, the managing agencies have tapped into the wealth of relevant expertise found in the community, thus ensuring that World Heritage has '... a function in the life of the community ...' (Article 5 of the Convention).

Acknowledgement: The authors thank all the GBRMPA officers who have contributed to the ongoing efforts to engage local communities over the years. Thanks also to Gail Barry, Nathan Walker and Peter McGinnity who contributed to a draft of this case study; also to Holly Savage, Wendy Kimpton, Gaye Collins, Carolyn Luder and Doon McColl for collating photographs.

3

Living World Heritage: Škocjan Caves, Slovenia

VANJA DEBEVEC[1]

The wonders of a karst landscape

Karst (Kras in Slovenian) is a rocky region in the south-east of the country, where numerous caves and chasms have formed in the limestone base. This karst underground system is unique in the world; here, natural values conditioned the development of cultural heritage in the past. The Škocjan Caves were inscribed on the World Heritage List in 1986. The Government of the Republic of Slovenia established the Škocjan Caves Regional Park through an Act of Parliament and designated the Škocjan Caves Park Public Service Agency as the managing authority of the protected area in 1996. Conservation and development of the surface and underground ecosystems are guided by legal provisions with the necessary restrictions on human habitation and interventions.

The underground course of the Reka River in the Škocjan Caves was included on the list of wetlands of international importance of the Ramsar Convention in 1999. The Škocjan Caves Park was included as Karst Biosphere Reserve in the world network of biosphere reserves of UNESCO's Man and the Biosphere (MAB) programme in 2004. The karst landscape includes dry karst meadows, rocky landscape, corrosion fissures, karren surface features, pastures, collapse dolines, dolines, cave system, underground canyon, the torrential Reka River, sinkholes, speleothems (mineral deposits), underground halls, canals and natural bridges. The underground river canyon in the Škocjan Caves system is a wetland of international importance under the Ramsar Convention.

[1] Head of Department for Research and Development, Park Škocjanske jame, Slovenia.

The Reka River in Škocjan Caves. The system of subterranean passages of the caves, fashioned by the Reka River, is a dramatic example of large-scale karst drainage.

Škocjan Caves formed in early Cretaceous limestone strata at the juncture with impermeable Eocene flysch. The waters of the surface flow of the Reka River gather at the base of Mount Snežnik. The Reka, with its 45 km surface flow, has carved a canyon in limestone before the Škocjan Caves. Under high, sheer rock faces, the river sinks into the Škocjan Caves at 317 m above mean sea level, runs through the Mahorčičeva Jama and Mariničeva Jama, resurfaces in the Mala Dolina collapse doline (Little Dolina) and reaches the lake in the Velika Dolina (Great Dolina). There it finally sinks into the Šumeča Jama (Murmuring Cave), 250 m in length and a maximum of 80 m in height with a volume of 870,000 m^3 and surface area of 16,700 m^2. It is followed by the Hankejev Kanal, a canyon 95 m high and up to 15 m wide. The Martelova Dvorana is the largest underground hall in the Karst at 308 m in length, 123 m in width, average height of ceiling 106 m, the highest point being 146 m, and with a volume of 2.1 million m^3. From there the river runs towards a siphon and into the Kačna Jama near Divača, before finally resurfacing as a tributary in the springs of the Timavo River in Italy.

The entire protected area lies in the Municipality of Divača. The transitional area, which was formed with the Karst Biosphere Reserve, measures 14,780 ha and covers the entire territory of the Municipality of Divača, which has a population of 25,000. The hinterland consisting of four municipalities – Ilirska Bistrica, Pivka, Postojna and Hrpelje-Kozina – covers 45,000 ha and has 80,000 inhabitants. The legal mandate of the World Heritage management authority is restricted to activities affecting the quality of water in the Reka River.

The area where the tranquility is only interrupted by dripping of water from the ceiling is called Tiha Jama (Silent Cave). Well-preserved stalagmite formations may be observed there. Velika Dvorana (Great Hall) in this part is distinguished by huge stalagmites, also named giants, that are up to 15 m high. The entrances to the caves are extremely picturesque collapse dolines. Their very name tells that they formed after the collapse of cave ceilings. Collapse dolines often have steep or vertical rock faces. In the Mala Dolina and Velika Dolina, through which the Reka River runs underground, an exceptional, geomorphologically and micro-climatically conditioned ecosystem developed and was preserved. Here, Mediterranean, sub-Mediterranean, Central European, Illyrian and Alpine floral elements occur side by side. Consequently, glacial and thermophile relicts, which are living testimony of past climatic periods, can be found here. Velika Dolina is the typical locality of the endemic *Campanula justiniana Witasek*.

In the underground system of the Reka River rare cave fauna are preserved. Among subterranean fauna the most numerous are copepods such as *Elaphoidella slovenica Wells* (syn. *Elaphoidella karstica*), a Škocjan Caves endemic. Copepod species *Moriaropsis scotenophila* and *Speocyclops infernus* were first described here. The olm (*Proteus anguinus*), a Dinaric endemic species of blind amphibian, is also found in the underground river.

The mysterious disappearance of the river underground and numerous surrounding caves held great significance for people, even in antiquity. In a relatively small area, in the immediate vicinity of the village of Škocjan and of the sinkhole of the Reka, as well as in the Vremska Valley to the east, there are over thirty archaeological sites, most of them in the caves. These are mainly cave posts, fortified settlements and burial grounds, which bear testimony to human settlement from the Mesolithic, Neolithic and Eneolithic periods, the Bronze and Iron ages to Classical antiquity, the Migration Period and the Early Middle Ages. Numerous burial grounds in the Mušja Jama and Skeletna Jama testify to the fact that this place was considered an important sanctuary of supra-regional significance. Rich archaeological finds bear witness to pilgrimages of people from the Mediterranean region.

Scientists began to systematically study karst features here and then spread their findings throughout the world. The Škocjan Caves region draws its outstanding universal value from its exceptional importance for the fundamental research of karst and karst features that has been going on since the 19th century. Many Slovenian words inform speleology and international geomorphologic terms such as 'karst' and 'doline'.

The mysterious disappearance of the river underground and numerous surrounding caves held great significance for people, even in antiquity.

32

Local people building a drystone wall, characteristic of the site.

Valuing people living with World Heritage

Škocjan Caves Park comprises 413 ha of the narrow protected area with a population of seventy people living in three villages, Škocjan, Matavun and Betanja. Stone houses, which were first thatched and later had stone roofing, are a feature of these settlements located above the caves. The layout of the village of Škocjan is typical of a hill-fort. In the past people built drystone walls, which protected and delimited their fields. The karst ponds are part of the local biodiversity and important in the local water network that served to collect rainwater for cattle. They have become a new habitat for numerous plant and animal species and contribute to the mosaic appearance of the cultural landscape.

The local lifestyle was largely forged by nature with its scant resources providing for survival. Additional income was generated working with explorers and managers of the caves, especially through assisting in the exploration of unknown parts of the caves and building access paths. Respect for ancestors and caves contributed to local communities developing a high awareness of the importance of World Heritage.

The World Heritage site is distinguished by an exceptional vulnerability of complex ecosystems. The ground with its limestone structure lets the water leak underground and with it also potential pollution. This can lead to the pollution of extensive pools of groundwater under the karst, which could destroy sensitive subterranean fauna as well as jeopardizing the quality of the water, which is part of the system of karst groundwater as a source of drinking water. In the past seventy years industrial development in Ilirska Bistrica has caused a dramatic decline in the quality of water of the Reka River. Appreciation of the region's outstanding universal value has however helped to ensure that the water is clean again, habitats are preserved and several species have reappeared. Past experiences of negative impacts and new consciousness of values has made the people living along

33

the river more attentive to conservation with a commitment to fostering a healthy environment.

The early development of tourism and the inclusion of local residents in this industry favoured the development of the local economy. Sustainability in tourism is certainly a challenge, but for the locals it is well worth it because they are aware of the great added value of trades performed according to the principles of sustainability. A few years ago the park revived the cave festival called Belajtnga that had died out in the 1940s. On the festival day, the last Sunday in May, locals and cave managers present their conservation work and organize guided tours. The festival has become a joint activity of the park management and the community to promote local production, encourage the use of local resources and the revival of traditional methods and customs. The branding of the park as World Heritage has consolidated the network of stakeholders contributing to the sustainable development of tourism.

The park has also developed an educational programme encouraging intergenerational transmission of local knowledge. Documentation and revitalization of heritage is contributing to children learning from their grandparents. This has become significant for the renovation of drystone walls that characterize the cultural landscape. Collective memory informs local people's identity and the functioning of the community. This has always been linked to cave exploration and tourism. The continuation of traditions is reflected in the growing interest for the production of typical foods made from local ingredients and in offering hospitality and experiences to visitors in renovated homesteads. Family-owned businesses drive the development of sustainable tourism. The past practice of including locals in explorations and maintenance works has contributed to the sense of ownership of the World Heritage by the community.

> On the festival day, locals and cave managers presented their conservation work and organized guided tours. The festival has become a joint activity of the park management and the community to promote local production, encourage the use of local resources and the revival of traditional methods and customs.

Responsible behaviour as management

The Škocjan Caves Park Public Service Agency is funded from the state budget that is supplemented with resources from ticket sales, international projects and donations. The Agency's Council, the supreme body of the park, includes representatives of the village community, the Municipality of Divača, the Ministries of Agriculture, Environment and Culture and the Slovenian UNESCO Office. Park management is informed by specialists from the fields of protected area management and speleology as well as through dialogue among stakeholders, especially local communities, park management and municipal authorities. Consciousness-raising and education of the public are crucial.

The site, located in the Kras region (literally meaning karst), is one of the world's best known for the study of karstic phenomena.

A few years ago, the community from Dane pri Divači suggested including the area above the Mejame cave within the protected area. Keen to protect the habitat of the olm (*Proteus anguinus*) they enthusiastically joined in the park's activities. They also established the Mejame Tourist Society, which was a developer and partner in projects for the renovation of karst ponds and construction of tourist pathways.

Škocjan Tourist Society promotes the site through summer events in the square in Škocjan and a tour around the Reka River sinkhole. Local women are active in the everyday work of the park and contribute to hospitality and banquets for important events held there. Children perform in cultural programmes at these events. A partnership with primary schools through the regional Network of Schools has been established.

Benefits through binding promises

35

The managing authority of the park under the Škocjan Caves Regional Park Act implements five-year programmes approved by the government. The primary task is to monitor the situation in ecosystems on the surface, in the caves and in the water of the river and other waterways. In the past ten years two water-monitoring stations have been set up on the Reka River. An automatic meteorological and precipitation station has been established in association with the Slovenian Environment Agency for the quality monitoring of weather data. The information generated is communicated to the public, especially through the daily broadcast on Slovenian radio, and has proved valuable to the locals in planning their daily activities.

Improved water quality has also led to an increase in the population of bats living in the Škocjan Caves. Restricting hikers to pathways allows birds to nest undisturbed on the faces of collapse dolines. By mowing dry karst meadows, the biodiversity of karst grasslands is maintained.

A special incentive for the locals to preserve cultural heritage are the funds that the park management grants annually to the protected area residents. After a public call for applications, the funds are granted for the renovation of

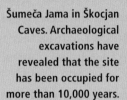

Šumeča Jama in Škocjan Caves. Archaeological excavations have revealed that the site has been occupied for more than 10,000 years.

buildings and cultural landscape and implementation is monitored by the park's expert commission. Work sessions, meetings and lectures are regularly organized to explain professional requirements in the protection of World Heritage and to raise public awareness of outstanding universal value. Training is also given to both park residents and the general public where local skills are identified in culinary traditions, medicinal herbs and other uses of natural resources.

Through the inclusion of the local community in the monitoring system, a network of trained volunteer environmentalists has been established. Today there are thirty-five volunteer environmentalists. The goal is to introduce participatory monitoring and the first attempts at this were made with primary schools. With the help of the internet, the schools record data about weather and tree species. This type of inclusion has become important in consciousness-raising and increasing social inclusion to strengthen local networking.

Training is also given to both park residents and the general public where local skills are identified in culinary traditions, medicinal herbs and other uses of natural resources.

The opening of the park's library has made it possible to keep audiovisual records of traditional customs and stories that are typical of the locality and are today only known to older people. In addition, the new meeting and promotion centre will host an information point with details of new public calls, data from the area of influence, and tourist service providers.

Twelve international projects have been completed to date. Project funds have been invested in the park and its surroundings for the renovation

Lake in Velika Dolina. The development of tourism has benefited the local community.

of significant remains and infrastructure on the surface and in the caves. Renovated buildings are also defined in terms of content so that they bring a new activity into the life of the locality. In this way we renovated the J'kopin Barn, which houses an ethnological collection about wheat and flour production. There were forty-five mills on the river in the early 19th century. Jurjev Barn today hosts an exhibition about the history of the exploration of Škocjan Caves.

Gamboč Homestead is the seat of the park management and Delez Homestead hosts a natural history centre with geological, biological and archaeological collections. Funding from European Structural Funds was used to renovate the path leading to the Mahorčičeva Jama and Mariničeva Jama as well as the tourist infrastructure, which will be further upgraded with future projects.

Beehouse in Škocjan Caves Park. Bee-keeping is one of the activities in the people's daily use of local resources.

As a result, today visitors can follow the river on its way underground from the first sinkhole and experience the history of the locality in Nanet's Homestead. The new meeting and promotion centre will bring together locals, who will offer their products and services, and foreign visitors or experts who will attend professional training in speleology, protected area management and natural and cultural heritage.

Experiential heritage tourism is further strengthened with clean water and air, absence of urban noise, and unspoilt nature in an exceptionally beautiful landscape. World Heritage branding and the quality of work in safeguarding outstanding universal value through the local communities in the park are promoted and appreciated internationally. The number of visitors has been growing steadily, exceeding 100,000 visits per year from 2008. The park management strives for personal contact with visitors and therefore the tours of Škocjan Caves are guided. But visitors can also experience the Mahorčičeva Jama and Mariničeva Jama and the Mala Dolina and Velika Dolina collapse dolines and the Škocjan Education Trail on their own. Now there are three homesteads in the park that offer food and lodging, and another two are being renovated.

Stakeholder cooperation informs efforts for positive solutions to the development of the locality and explores possibilities for a better quality of life for both residents and visitors. Activities are planned emphasizing the interconnectedness of nature and culture as an integral part of the site's outstanding universal value.

Sharing through learning

In 2000 the Škocjan Caves Park established an international network of schools that includes five from Slovenia and two from the Italian side of the border. The founding document comprises activities and goals, among which are popularizing natural and cultural heritage, cultural differences, strengthening individual

In 2000 the Škocjan Caves Park established an international network of schools that includes five from Slovenia and two from the Italian side of the border. Through the network of schools the park also connects with sites from other countries.

identities in space, and responsible behaviour. Through the network of schools the park also connects with sites from other countries.

The park develops project assignments for primary schools and enables them to take part in other parks' international projects. This encourages the exchange of experiences between mentors and pupils and the acquisition of additional expert knowledge, from parents to decision-makers at municipal level. The purpose of practical research work performed by pupils in schools and in the park is to apply knowledge and to confirm their own solutions to environmental problems outside schools. So, on the basis of the concept of sustainability and the importance of natural science in its application in educational programmes, the main principles of education for sustainable development as a system of values are embedded in the adopted formula L × 3L: *Learn to Love, Learn to Live, Learn to Last.*

Creative development through World Heritage

Conserving World Heritage means more than protected areas and values. It is a form of development resulting in quality living. In accordance with the World Heritage Convention, the park created an educational programme that makes access, information and knowledge of the outstanding universal value available to everybody, enhances visitor experiences of environmental phenomena, and promotes individuals' responsible behaviour and conduct in the karst area. Future plans include improving the active network of stakeholders together

39

The park created an educational programme to make information about the outstanding universal value of the site accessible to all.

with the network of schools and faculties, further contributing to economic and social development.

Living with World Heritage means protecting natural and cultural resources but also creating new heritage values. Through a constant search for new knowledge, through respectful learning from traditions, we can find new solutions for life and work in the future. With respectful and persistent work we can create economic, social and ecological conditions for the welfare of people, for quality living. And let us not forget the power of words with which we will greet future generations:

'Many simple words are needed, like bread, love, goodness,
lest we stray from the right path, blind in the darkness.'

(Tone Pavček)

4

Challenges of protecting island ecosystems: Socotra Archipelago, Yemen

KAY VAN DAMME[1]

'Galápagos of the Indian Ocean'

The Socotra Archipelago World Heritage site (Yemen) is situated in the Western Indian Ocean, a 250 km long archipelago about 100 km east of the Horn of Africa and 380 km south of the Arabian Peninsula. From west to east, the archipelago is made up of the islands Abd al Kuri (133 km²), 'The Brothers' Samha (41 km²) and Darsa (17 km²) and finally Socotra, the easternmost and largest island of the group, covering some 3,625 km² of land surface (Cheung and DeVantier, 2006; Banfield et al., 2011). Socotra is about 1,550 m at its highest point in the Haggeher mountains and most of its surface consists of elevated limestone plateaus, bordered by coastal areas and a central depression.

The island of Darsa and a few barren islets are uninhabited by man, yet serve as important sites for large populations of Socotra Cormorant and other seabirds. Despite its proximity to Africa and the fact that it is biogeographically considered a part of this continent, the archipelago lies on a micro-continent (the Socotra Platform), which geologically belongs to Southern Arabia and which has been separated from it by a deep sea (the Gulf of Aden) for at least 18 million years (Cheung and DeVantier, 2006; Van Damme, 2009). Like the Seychelles, the base of this archipelago is made up of ancient granite that once belonged to Gondwanaland. It contains a remarkably rich natural and human history. Sometimes referred to as the 'Galápagos of the Indian Ocean' (Sohlman, 2004), the Socotra Archipelago is, in contrast to its Ecuadorian

[1] Chair of the Friends of Soqotra and Researcher at the College of Life and Environmental Sciences, University of Birmingham, United Kingdom.

Fisherman and boats at Noged beach, south Socotra Island. Socotra faces challenges in the marine and terrestrial ecosystems which affect the food security. In recent years fishing has shifted from low- to high-impact.

42

counterpart, continental in origin, therefore much older, and the origin of its biota has a more complex history (Van Damme, 2011).

In its human history, the main island has functioned for several centuries (Greek and Roman times) as a well-known major trading point for natural products such as aloe, dragon's blood resin, myrrh, frankincense and ambergris. Socotra is also the largest island in the Arab world, *The Pearl of Arabia*, known to many for its biodiversity and therefore as significant to the region as Madagascar is to Africa (Van Damme, 2011). The island is of considerable importance to Yemen, a natural oasis in a country plagued by poverty, playing a symbolic role in maintaining nature awareness as part of the national culture. The iconic Socotra Dragon's Blood Tree is a national emblem, imprinted on the Yemeni 20 rial coin.

Socotra was inscribed on the World Heritage List in 2008, for its biological diversity and threatened species ('contain the most important and significant natural habitats for in-situ conservation of biological diversity, including those containing threatened species of outstanding universal value from the point of view of science or conservation'): 'Socotra is globally important for biodiversity conservation because of its exceptional level of biodiversity and endemism in many terrestrial and marine groups of organisms' (IUCN/UNESCO, 2008). With a core area of 410,460 ha (68 per cent terrestrial, 32 per cent marine), it is now one of the larger insular natural World Heritage sites. The diversity of ecosystems is well represented within the nominated areas, which cover 73 per cent of the terrestrial surface of Socotra, around 50 per cent of the coastal area, and all the surface and coasts of the smaller islands and islets, a total of twelve terrestrial and twenty-five marine protected areas (IUCN/UNESCO, 2008; UNEP/WCMC, 2008). The archipelago counts 835 higher plant species of which 37 per cent are unique; when the number of endemic species per km² is taken into account, this relatively small archipelago even ranks in the top five of botanically richest continental islands in the world

Youth holding fish at Hadiboh, Socotra Island. Overfishing leads to less diversity and a shift from larger to smaller fish.

(Banfield et al., 2011). It also has about 90 per cent of the reptile species and 95 per cent of the land mollusc species unique to the archipelago (sources in Cheung and DeVantier, 2006; Van Damme and Banfield, 2011). More than 730 fish species and 283 coral species present the extensive coral reefs that border these islands, significantly more than in the larger Galápagos (IUCN/UNESCO, 2008). Many of the plants are now being screened for medicinal properties, which have been used by local people for centuries, leading to a vast ethnobotanical knowledge that is deeply embedded in the local language (Morris, 2002; Miller and Morris, 2004). Research has confirmed that several of the Socotran traditional medicinal plants have strong antibacterial and antioxidant properties (e.g. Mothana et al., 2009), which illustrates the significant scientific value of the archipelago's indigenous species (Van Damme, 2011). The remarkable ecosystems have led to further international recognition of Socotra: as a UNESCO Man and the Biosphere Reserve (2003), WWF Global 200 Terrestrial Ecoregion, Plantlife International Centre of Plant Diversity and part of Conservation International's Horn of Africa Hotspot (e.g. Van Damme and Banfield, 2011; Scholte et al., 2011).

43

Many of the plants are now being screened for medicinal properties, which have been used by local people for centuries, leading to a vast ethnobotanical knowledge that is deeply embedded in the local language.

Socotra also has important geological features, in some areas containing karstic limestone formations similar to the Tsingy in Madagascar and harbouring extensive cave systems (De Geest, 2006). It has exceptional cultural features, such as the unique, unwritten language of the Socotri people that contains an invaluable link to the ecosystems (Morris, 2002; Miller and Morris, 2004). Its archaeological sites such as Hoq Cave contain ancient Indian scripts that provide keys to long-lost trade routes in the Indian Ocean, or the enigmatic rock inscriptions at Eriosh of which the age and origin are still a subject of debate (sources in Cheung and DeVantier, 2006). For geologists, archaeologists, linguists and anthropologists, the archipelago therefore contains many additional values besides the biodiversity. The Socotri people have maintained a relatively low-impact resource management for centuries (Morris, 2002; Scholte et al., 2011; Van Damme and Banfield, 2011), with no local extinctions in molluscs, reptiles and birds in the last century, in sheer contrast to most islands in the world (Van Damme and Banfield, 2011).

Management of Socotra

Its archaeological sites such as Hoq Cave contain ancient Indian scripts that provide keys to long-lost trade routes in the Indian Ocean, or the enigmatic rock inscriptions at Eriosh of which the age and origin are still a subject of debate. For geologists, archaeologists, linguists and anthropologists, the archipelago therefore contains many additional values besides the biodiversity. The Socotri people have maintained a relatively low-impact resource management for centuries.

The site is managed by the Environmental Protection Agency (EPA Socotra Branch) and by the Ministry of Water and Environment of Yemen. Before its nomination, Socotra biodiversity conservation had been developed through ICDPs (Integrated Conservation and Development Projects), that started in the late 1990s led by the UN (UNDP-GEF project 1997–2001, the Conservation and Sustainable Use of the Biodiversity of the Socotra Archipelago; Socotra Conservation and Development Programme, 2001–2008; see Elie, 2008; Scholte et al., 2011). The early projects were translated into the Socotra Conservation Zoning Plan in 2000, defining national parks and nature sanctuaries, providing the legal framework for conservation in the archipelago (Cheung and Devantier, 2006; Scholte et al., 2011) and the basis on which World Heritage properties were defined. The recommendation for the establishment of the Socotra Island-Wide Authority for more local management of the World Heritage site (UNDP, 2008; Scholte et al., 2011; IUCN/UNESCO, 2008), is yet to be implemented. UN projects for Yemen such as the Socotra Governance and Biodiversity Project (2008–2013), which focused on these governance issues and legal frameworks for biodiversity conservation on the island, were suspended as a result of the Arab Spring, during which most international donors retreated just a few years after the nomination of Socotra as a World Heritage site, resulting in impacts on ongoing conservation efforts (Tollrianova, 2011; Yahia, 2011; Van Damme, 2011). Since mid 2011, the archipelago has been promoted from local district to intermediate authority (Peutz, 2011) and a new sub-governor of Hadramawt was appointed in March 2012 for Socotra affairs.

A timely inscription

The World Heritage inscription in 2008 arrived exactly at a time when human-mediated impacts on the Socotran ecosystems took on an accelerated pace. This is a result of economic shifts and developments that started in the 1960s and culminated in a new wave of challenges that arrived in the late 1990s, when the island opened to the world (e.g. Cheung and DeVantier, 2006; Elie, 2004; 2008; Morris, 2002; Van Damme and Banfield, 2011). In 1966, before the start of tourism and when herding was still the core activity of the population,

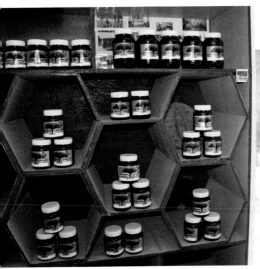

Socotra Honey Centre, Hadiboh, the Beekeepers Association. Example of community participation in protection of the biodiversity of the site by low-impact production of honey.

83 per cent lived in the hinterland and 17 per cent on the coast (and a seasonal switch had taken place over centuries, from fisherman to herder and vice versa; Morris, 2002), which later became about 60 per cent in the coastal areas and 40 per cent in the hinterland as a result of socio-economic shifts (census 2004; Elie, 2008). Now, the situation is exactly the reverse: for a population of some 50,000 people on Socotra, about 88 per cent, most of whom are immigrants attracted by development, trade and tourism, are concentrated in less than 10 per cent of the island's surface (in two coastal urbanized areas, Hadiboh and Qalaansiyah), which leaves 12 per cent, nearly all local residents, living outside these two towns. Such migrations and the socio-economic shifts on Socotra, the loss of traditional fisheries and transhumance, and the correlated local shifts in resource use, have strong impacts on the marine and terrestrial ecosystems and therefore on the heritage of future generations.

In the marine ecosystems, recent estimates of six marine sites on Socotra Island show a severe (a five- to sixfold) decline in mean biomass of consumable fish because of a shift from low-impact subsistence fishing to high-impact commercial overfishing, a drop from 1.53 t/ha to 0.24 t/ha and a decline of about 75 per cent in total abundances, over the years 2007–2011 (Zajonz et al., 2012a). In the terrestrial ecosystems, changes in traditional transhumance have resulted in an increase of grazing impacts on Socotra's limestone plateaux. The correlation between land degradation and the breakdown of traditional pastoralism has been clearly shown on the island, with 40 m^3 soil disappearing over a period of only three years in a single gully head (Homhil Protected Area; Pietsch and Morris, 2010). Although only outside monsoon periods (Scholte and De Geest, 2010), impacts of tourism have also increased recently. Tourist numbers have doubled every eighteen months since 2003 (Scholte et al., 2011), reaching around 4,000 foreign visitors in 2010 (Socotra Governance and Biodiversity Project; see Van Damme, 2011). The impacts are various, yet generally lead to an increase in tourism infrastructure, attraction of investors

45

and immigrants, and an increase of illegal collection and waste on the island, generally evolving from low-impact ecotourism to high-impact tourism (Van Damme and Banfield, 2011) as in many islands (e.g. Deidun, 2004). During a recent survey, when asked what visitors enjoyed most about their ecotour on Socotra, 58.6 per cent answered 'beaches' (the highest score), compared with 9 per cent answering 'fauna and flora', illustrating that ecotourism on the island is not always 'eco' (Mayer, 2009). Considering the recent speed of human-mediated ecosystem impacts shown in the few examples above, a loss of biodiversity through extinctions of the local fauna and flora and a sudden breakdown of the unique insular ecosystems with further effects on local communities, are inevitable (Van Damme and Banfield, 2011). This is where the UNESCO World Heritage inscription now plays a significant role – it provides a timely tool in emphasizing the importance of the island's biodiversity and the continuous need for the protection of the ecosystem services for future generations. The inscription has provided a strong motivation for more strategic approaches to the challenges confronting these insular ecosystems and their people.

Local initiatives

Local communities practised conservation until the 1960s through traditional resource use, such as in transhumance and fisheries, having an extensive set of traditional laws that ensured sustainable management (Morris, 2002). A shift from barter to cash economy disrupted this close link between the local inhabitants and their environment and this is probably irreversible (Elie, 2008). Various projects have however recognized the leading role of communities in conservation and development, resulting in a certain extent of local involvement (see Scholte et al., 2011). Local communities have expressed a wish for stronger involvement and role in the decentralization process (Peutz, 2012; Elie, 2008). There are several examples of public and private-sector initiatives and of community participation that help conservation, which started before the time of nomination and continue to date.

One example is the *ex situ* plant conservation at Ahmed Adeeb's Nursery near the capital Hadiboh (Van Damme and Banfield, 2011; Cheung and DeVantier, 2006). It started as a private family initiative in 1996, and was initially supported by the Ministry of Agriculture, EPA, the Socotra Conservation Fund and the UN-SCDP (Socotra Conservation and Development Programme). Since 2007, knowledge has been shared between research institutes (Edinburgh Royal Botanic Garden, UK, and Palermo Botanic Garden,

Local communities practised conservation until the 1960s through traditional resource use, such as in transhumance and fisheries, having an extensive set of traditional laws that ensured sustainable management. A shift from barter to cash economy disrupted this close link between the local inhabitants and their environment and this is probably irreversible. Various projects have however recognized the leading role of communities in conservation and development, resulting in a certain extent of local involvement.

46

Fishermen singing and dancing to celebrate a new find of a freshwater source for the village (at Irriseyl, Socotra Island). The Socotran people are proud of their unique, unwritten language.

Italy), training Mr Adeeb in maintaining the thousands of nursery seedlings and systematic data collection. The nursery contains over 15,000 seedlings of the island's symbol, the Socotra Dragon's Blood Tree, which provide a genetic stock of the populations if the species, as predicted, declines in the future (Attorre et al., 2007). Grown from just 38 plant species in 2007 to 117 species in 2011, the nursery holds 37 per cent of the unique higher plants of the archipelago (Van Damme and Banfield, 2011). At the same time, Adeeb's nursery functions as a private ecocamp where visitors can put up tents, providing an opportunity for awareness, research and education. The nursery may be considered as one of the more successful private initiatives on the island, illustrating how important the role can be of local initiatives in protecting the integrity of Socotra's outstanding universal value. Another example is the semi-private initiatives for rubbish recycling that emerged a few years ago in Hadiboh, where waste is a major problem (Van Damme and Banfield, 2011). Money is paid for the collection of plastic, which is cut and shipped to the mainland for recycling. Recently, in 2011, groups of local people spontaneously gathered up rubbish in the streets of Hadiboh, without a backing project or sponsorship (Peutz, 2012).

Other examples of private-public sector partnerships and community participation include the local NGO, Socotra Foundation for Bees Protection and Breeding. Bee-keeping started in 2004 with the SCDP and the local Agricultural Department jointly initiating a project with wooden beehives (see Cheung and DeVantier, 2006). These skills were shared with French professional bee-keepers, with previous projects training over 200 local bee-keepers who could raise their standard of living through the sale of honey. The initiative will be extended to a total of 110 families (GEF Small Grants Project 2011–2013) ensuring large-scale community participation with equal opportunities for both genders. The initiative also plays a role in biodiversity conservation through enhancing pollination of indigenous plants on Socotra,

47

A young Socotri walking on the beach near his home close to Terbak, north-eastern coast of the island. The sustainable development of Socotra depends on a balance between local needs and preservation.

therefore keeping communities involved in protecting the natural botanical environment.

Besides a number of private and public-sector initiatives including ecocamps and ecoguides, over a hundred NGOs are currently active on Socotra. Several Socotran Wildlife Conservation NGOs focusing on the importance of local biodiversity and nature awareness are now starting to take shape, strengthened by the fact that Socotra has received World Heritage status.

Where do we go from here?

World Heritage status has undoubtedly increased the attention paid to several major challenges, such as grazing, road development, invasive species and tourism development on the island, leading to more strategic approaches in tackling some of these issues and giving local communities and conservationists an important additional protection tool.

At present, with recent institutional reforms in Yemen and on Socotra (Van Damme, 2011; Peutz, 2012), World Heritage inscription provides a number of opportunities for future conservation of the natural and cultural heritage. World Heritage status has undoubtedly increased the attention paid to several major challenges, such as grazing, road development, invasive species and tourism development on the island, leading to more strategic approaches in tackling some of these issues and giving local communities and conservationists an important additional protection tool. Road development, a major concern on the island, has slowed down since the inscription. The issue of recent coral mining in 2012 on Socotra (Zajonz et al., 2012b), where tons of coral were taken from a marine core area and exported, is being addressed thanks to the intervention of UNESCO.

The World Heritage nomination has also stimulated a number of projects, case studies and research that form the baseline for future reference and a way to monitor the impacts of current and future challenges – all of which emphasize the important historical and current role of the resident population and involvement of local communities, yet sound a warning about the recent increase of unsustainable resource use (e.g. Van Damme and Banfield, 2011; Scholte et al., 2011; Zajonz et al., 2011).

A solitary Dragon Blood tree grows by the side of the road. The first paved road on the island was built by the Yemeni Government in 2005.

Threats to the integrity of the biodiversity of the Socotra World Heritage site are numerous (Cheung and DeVantier, 2006; Van Damme and Banfield, 2011; Zajonz et al., 2011). It is well known that island biotas have higher extinction rates than continental ones because of smaller areas of occurrence of species, a relatively higher proportion of unique species and higher sensitivity to impacts, which is the case on Socotra (Van Damme and Banfield, 2011). It is harder, from the viewpoint of capacity-building and local training, to protect island biodiversity. Although the tourism sector is potentially very important for local income and development, an increase in tourism can have severe negative impacts on an island that comprises 72.6 per cent of national park and where unique species are very localized, if not combined with proper management. As mentioned above, low-impact environmentally interested travellers (ecotourists) have been gradually replaced by higher-impact tourists over the course of a decade, treating Socotra as a beach, expecting increasing standards of luxury and with little income for the local communities (Mayer, 2009; Van Damme and Banfield, 2011). World Heritage status can hopefully increase awareness among visitors and decision-makers as well as the local communities of the future, of the importance yet also the fragility of the island's ecosystems, which include the local culture and language.

We can learn from insular ecosystems all over the world for further management of the Socotra site, and from past experience. A few general points may be considered important for the sustainable management of the World Heritage site (see Attorre et al., 2007; Peutz, 2012; Van Damme and Banfield, 2011; Morris, 2002; Pietsch and Morris, 2010; Miller and Morris, 2004; Cheung and DeVantier, 2006; Mayer, 2009; UNDP, 2003, 2008; Zajonz et al., 2011).

- An invaluable key to management is the protection of the unique local language of Socotra. A poetry contest on the island and incorporating the language into school curricula should be regarded at least as important to the conservation of Socotra's natural heritage as any initiative for strategic

49

Socotran desert rose, *Adenium obesum* subsp. *socotranum*. Socotra is globally important for its exceptionally rich and distinct flora and fauna.

biodiversity management. On the verge of extinction, the rich language retains exceptional links between the ecosystems and their people, critical for community involvement. The language can still retain a sense of collective care for the ecosystems by the true stewards of the archipelago, the Socotri themselves. More than the environment, the Socotri people consider the language as the basis of their cultural identity and pride, which is important (e.g. Peutz, 2012).

■ Controlled tourist management is important to the future of the site, and limitations to the number of tourists, the type of tourism and the infrastructure should be determined for the future. Currently, Socotra is at a point in time of tourism impacts comparable to the Galápagos in the 1970s, which had severe environmental consequences. More revenues to local communities from ecotourism should be a priority (e.g. Mayer, 2009).

■ The efforts of increased local authority management of the site require local involvement in decision-making. Ideally, organizations work together as a whole to face the main challenges to the site, sharing a single community ideal: continuing protection of the unique heritage of Socotra. Further efforts in preparing a legislative framework for protection, drawing on

Building new tourist accommodation. Tourism has improved in recent years, with visitor awareness of biodiversity conservation.

51

traditional knowledge and involvement of local people in conservation through training and participation, are important.

- It is vital, for the future, to prioritize efforts that focus on protecting what can still be protected. Present-generation knowledge bearers of traditional practices that contributed to conservation on Socotra for centuries are still active. The intergenerational transmission of their knowledge could still minimize the negative impacts on the local environment as they have just begun. Efforts towards protecting the archipelago against well-known challenges for island biodiversities, such as invasive alien species, habitat and land degradation, as well as waste management, retaining local culture in future education, the impacts of climate change, resource management (water, fisheries, wood) and renewable energy, are important directions for the future.

A new momentum

In conclusion, the World Heritage site is playing a significant role and contributing to the sustainable development of Socotra. The nomination has been a positive and necessary evolution for the long-term preservation of the archipelago. World Heritage status has created a new momentum that allows people, both locally and internationally, to recognize Socotra's natural and cultural values over its beauty and short-term use and provides a much-needed tool to deal with the challenges that its people and the environment will face in the coming centuries.

The World Heritage inscription provides a baseline for all future initiatives to recognize and protect the rich biodiversity of the Socotra Archipelago. In addition, it allows the Socotri people to retain a sense of pride and care of their heritage for the long term, at a time when that identity and the natural richness are at risk. It is hoped that, in a century from now, Socotra will be largely the same, with intact ecosystems, without extinctions and with poets that sing about the majestic Haggeher mountains that will outlive us all.

5

Cultural landscapes: challenges and possibilities: Vegaøyan – The Vega Archipelago, Norway

RITA JOHANSEN[1]

Ancient landscape

The Vega Archipelago was inscribed on the World Heritage List in 2004 as the first Norwegian cultural landscape. The archipelago is a shallow-water area just south of the Arctic Circle on the west coast of Norway – an open seascape and coastal landscape made up of a myriad of islands, islets and skerries. This cluster of low, treeless islands centred on the more mountainous island of Vega is a testimony to people who developed a distinctive and frugal way of life in an extremely exposed seascape.

Fishermen and hunters have lived on the island of Vega for more than 10,000 years. As numerous new islands gradually rose from the sea after the last Ice Age in Europe, the characteristic landscape became shaped in the interplay between fishermen-farmers and a bountiful nature in an exposed area.

The unique tending of eider ducks was a central part of their way of life. People built shelters (houses) and nests for the wild eiders, which came to the islands each spring. The birds were protected from all manner of disturbance throughout the breeding season and became gradually semi-domesticated. In return, the people would gather the valuable eider down and make duvets, when the birds left their nests with their chicks.

[1] Site coordinator, Managing Director, Vega World Heritage Foundation, Norway.

53

The Vega Archipelago consists of 6,500 islands, islets and skerries.

As early as the 9th century, tending eiders was reported to be one way that people in Norway made a living, and the Vega Archipelago was the core area for this tradition. The women played a key role, so World Heritage status also celebrates their contribution to down production.

As early as the 9th century, tending eiders was reported to be one way that people in Norway made a living, and the Vega Archipelago was the core area for this tradition. The women played a key role, so World Heritage status also celebrates their contribution to down production.

The tradition remains alive today, albeit to a smaller extent than previously. Each spring, the bird tenders live on the islands for two months to look after the eiders in the breeding season. Afterwards, they clean the down and make exclusive eiderdowns and other products linked to the down tradition.

People no longer live all year round on the small islands in the Vega Archipelago. The approximately 1,300 inhabitants of the borough of Vega live in the buffer zone. This is on the main island of Vega, a gateway to the World Heritage area. Farming is the most important occupation there, but there are also some twenty-five fishermen, and other people work in the service, petroleum and small-scale tourism industries.

Local initiative and ownership

Sustainable social, environmental and economic development of the Vega Archipelago World Heritage site and its buffer zone requires good management,

A woman sitting close to a nesting eider duck. The eiders become semi-domesticated during the breeding season.

awareness and dissemination in keeping with the obligations that accompany the World Heritage status.

The nomination process started as a local initiative based on the Nordic report, World Heritage List in the Nordic Countries, NORD 1996:30. This report had suggested considering the 'Northern Norwegian islands' as one of four new areas in Norway for nomination.

Consequently, from the outset, the bird tenders on the islands, local farmers and other stakeholders were positive and keen to help to look after the World Heritage area.

Almost the whole of the World Heritage site is privately owned and many people feel they are an integral part of it and hence expect to be included in the work and the decision-making. Furthermore, there are other stakeholders who have a strong connection to the area and they also want to be an active part of its management. The communication between the management authorities, coordinating bodies and those with local interests at stake is therefore important. After the Vega Archipelago received World Heritage status, the work has been characterized by close cooperation and a good dialogue between the local community and the management authorities. Such dialogue also helps to engender awareness and an understanding of why the area must be preserved for posterity through sustainable development.

Cooperative management

Landowners, authorized users, the Vega Borough Council, Nordland County Council and the national public authorities work closely in unison to preserve the cultural landscape of the Vega Archipelago with its cultural and biological values. The Norwegian Government allocates funds annually to carry out management, information, restoration and local value creation efforts.

A positive effect of this is increasing numbers of grazing livestock, and haymaking on more and more areas, which are helping to restore overgrown land and safeguard the mosaic in the landscape. Also there are more bird tenders and more eider birds in the Archipelago than before the inscription.

55

Management plans for the World Heritage site have been drawn up in cooperation with local stakeholders, based on documentation of bygone practices and mapping of the existing biological diversity. The outstanding universal value of the site is being documented and passed on to the local community and visitors by teaching children and young people through 'hands-on' projects, research, guided excursions and information via the internet, brochures and the like. The association Friends of the Vega Archipelago is helping to pass on traditional knowledge gained by experience.

The Municipal Plan for Vega also contains a strategic part and a land-use part that ensure development in other parts of the World Heritage site and the buffer zone, safeguarding the outstanding universal value. The strategic objective of the plan is that Vega is to be developed as an international centre for knowledge on management, dissemination and sustainable use of the natural environment and cultural landscape on the coast.

Organization

The local school has a World Heritage plan, including 'hands-on' projects where the pupils go out into the site to help to construct 'houses' and nests for the eiders, make hay and clear rubbish which drifts ashore on the ocean currents. They also gather information on the intangible cultural heritage of the area.

The Vega Archipelago World Heritage Foundation was set up to promote the site and coordinate the local World Heritage work. The Foundation has a council composed of representatives of national, regional and local governments: the mayor of Vega and another representative from the borough council, the Ministry of the Environment represented by the Directorate for Nature Management, the county governor of the region, the director of the regional museum and two elective members.

The Foundation also has a cooperative board made up of representatives of eighteen local NGOs and associations, all of which regard themselves as stakeholders in the local World Heritage work. They include for example associations of farmers and fishermen, Friends of the Vega Archipelago, a small-scale business association, a tourism association and an ornithological society.

At the administrative level, a Team World Heritage Vega, made up of representatives from Vega Borough Council, Vega Archipelago World Heritage Foundation, the Norwegian Nature Inspectorate and Visit Vega/The Tourist Information, meets monthly.

World Heritage in young hands

The UNESCO Special Project, Young People's Participation in World Heritage Preservation and Promotion, launched in 1994, gives young people a chance to voice their concerns and become involved in protecting the world's natural

The Vega Archipelago habitats. There is evidence of human settlement from the Stone Age onwards on the archipelago.

and cultural heritage. It is important to engender a sense of ownership and belongingness in the coming generation that will carry the World Heritage work on into the future.

The Vega Archipelago World Heritage Foundation therefore puts high priority on raising awareness among children and young people and on cooperating with them. Local and regional schools play an active role in this work. The local school has a World Heritage plan, including 'hands-on' projects where the pupils go out into the site to help to construct 'houses' and nests for the eiders, make hay and clear rubbish which drifts ashore on the ocean currents. They also gather information on the intangible cultural heritage of the area.

A website set up for children and young people has information about all these activities (www.ungehender. no). It is an important tool for young people to disseminate the outstanding universal value and through this build an understanding of why the archipelago has World Heritage status and how they might participate in the safeguarding of the site.

Vulnerable nature and tourism

The birdlife and natural environment in the Vega Archipelago are in general vulnerable. The infertile landscape tolerates little trampling by visitors. The practice of tending the eiders and the birdlife requires peaceful conditions in the breeding season.

The Foundation puts great priority on raising awareness of the outstanding universal value and on visiting the area on nature's own terms. An impact assessment for controlled traffic in the World Heritage area has been performed to ensure the sustainable development of the area. The Foundation and the Norwegian Nature Inspectorate have joined forces to prepare maps, information boards and other information material with the intention of ensuring that boat people and other visitors to the area know how to behave.

The local community has also collaborated with regional and national authorities to develop a strategy for tourism whose keywords are target development, orderly traffic and local value creation. The majority of undertakings concerned

The Vega Archipelago forms a cultural landscape of 103,710 ha, of which 6,930 ha is land.

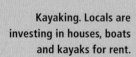

Kayaking. Locals are investing in houses, boats and kayaks for rent.

with providing information about the World Heritage status and site, excursions and infrastructure are in the buffer zone. This also applies to making paths, arranging guided walks and hiring out kayaks and bicycles, so that tourism in the World Heritage area itself can take place in a safe and controlled manner.

Sustainable economy

The Vega Archipelago World Heritage site is located far from the large concentrations of population in Norway. No one lives in the area throughout the year and there are no scheduled sailings. It is difficult to make the small-scale tourism activities on the islands profitable. Vega Borough Council has therefore been cooperating with the Norwegian Directorate for Cultural Heritage and regional authorities on the programme Creating New Assets in the Cultural Heritage Sphere. This has resulted in restoration of much of the built vernacular heritage and the buildings have been brought into use in connection with tourism. An important objective for the years ahead is to establish new projects based on the cultural heritage, in cooperation with the landowners and public authorities.

Craftworkers restoring buildings. A special competence programme trains local craftsmen in traditional restoration methods.

Vega is, moreover, one of five Norwegian pilot destinations for sustainable tourism. Sustainable tourism development requires the informed participation of all relevant stakeholders and the strategy in Vega has been to have a political debate and to ensure wide participation and consensus building. The goal of the programme is to develop strategies and activities for sustainable tourism and preservation of the cultural heritage and nature through awareness, development and enhancement of the historic heritage, authentic culture, traditions and character of the community.

The programme is also intended to help to improve social and economic values. An effort is being made locally to develop new activities and provide accommodation and food under the management of local owners and using local resources. It is also important to increase the number of visitors. When the archipelago was inscribed on the World Heritage List in 2004, Vega had some 5,000 visitors each year. This figure has now risen to 30,000 and efforts

Fishermen with their catch. The local way of fishing on a small-scale basis is an important factor in the inscription of this cultural landscape.

are being made to increase the number to about 50,000 to make the tourist industry more profitable. Almost all of these visitors will stay in the buffer zone. Only a few hundred visit the actual World Heritage area in the archipelago in the course of the summer.

Local benefits of inscription

It was the local community that took the initiative to have the area inscribed on the World Heritage List, seeing the status as a means of protecting the outstanding universal value and encouraging sustainable development in a depopulated area with few possibilities for young people to find work. A great deal has changed following the inscription. The development has been based on local commitment and on prioritizing the World Heritage site in national budgets and programmes. First and foremost, the preservation of the cultural heritage and the local development has been strengthened during the eight years since the inscription of the Vega Archipelago.

■ The eider down tradition has been revitalized. Work has resumed on several of the old down islets. The number of bird tenders and the quantity of down have risen. The tenders now receive compensation enabling them to stay on

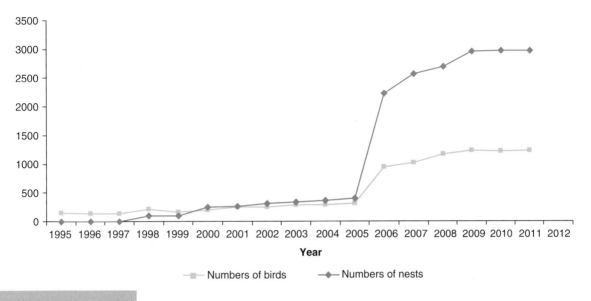

Eiders and nests on all the down islets

the islands in the breeding season, which also leads to new tenders being recruited. The tenders have started the Nordland Eider Duck Association, which arranges courses for tenders who can keep alive the ancient traditions associated with the old down islets and for other interested parties.

■ Funding is available to protect the cultural landscape, and the managed area is increasing year by year. At the time of inscription, about 400 sheep and

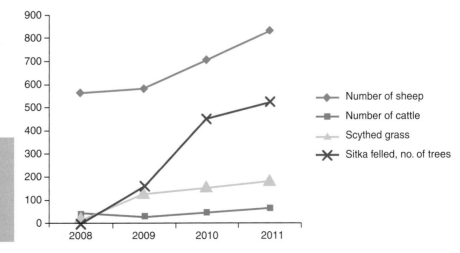

The managed area of the cultural landscape is increasing year by year.

Youth haymaking. Pupils participate in the management of the area.

20 head of cattle were grazing the archipelago. In 2011 there were 844 sheep and 70 head of cattle. Likewise, haymaking is taking place over an increasingly larger area; some 17 ha in 2011 as opposed to 2 ha when the inscription took place in 2004. Overgrowing is therefore declining and biodiversity is increasing.

- Important elements in the built vernacular heritage have been restored and are being put to new uses in both the World Heritage site and the buffer zone. More than twenty projects in the value-generating programme, Creating New Assets in the Cultural Heritage Sphere, have been carried out.

- There is local innovation and development, and local people are investing in small-scale tourism. Several tens of millions of Norwegian kroner have been invested in the local community since the Vega Archipelago received World Heritage status.

- Vega is building networks with other World Heritage sites. Cooperation with sites in Europe and Africa is providing valuable knowledge on management, how to impart information and the use of cultural and natural environments in World Heritage sites.

Development in all the above areas is promoting social, cultural, environmental and economic value creation.

Lessons we have learned

Heritage in general represents items of outstanding and special value: sites and objects which we seek to protect and conserve. Cultural and natural heritages are vital parts of our identity and world outlook. The World Heritage status, with the preservation and promotion of the outstanding universal value, has meant that the identity and pride of the people of Vega have been strengthened, which is important for the development of a local community. This pride has been based on involvement and local cooperation. In cases where

Studying seaweed and shells. Local schoolchildren in the region have science lessons at the World Heritage site.

we have failed with the information and collaboration, we experience difficulties in the implementation of the Management Plan and awareness of values. Thus, when the plan is updated in 2012, it will be in close collaboration with a reference group of local stakeholders, securing their commitment through discussions and common goals for the site management in the years to come.

World Heritage status is swelling the flow of visitors wanting to visit the islets, most of whom do not always behave appropriately in a vulnerable nature landscape. But at the same time they are engaged with the World Heritage site and they want to participate in its protection. Therefore the Foundation has a strategy for protection through information, education and involvement. A World Heritage Visitor Centre on Vega is being planned, which will provide information about UNESCO and the World Heritage Convention, the outstanding universal value of the site, but also the attractions of the area. The centre will also be the starting point for guided trips to the islands.

63

> World Heritage status is swelling the flow of visitors wanting to visit the islets, most of whom do not always behave appropriately in a vulnerable nature landscape. But at the same time they are engaged with the World Heritage site and they want to participate in its protection.

Safeguarding the outstanding universal value

Providing comprehensive information and communicating the site values and need for protection will continue. This includes developing local expertise and resources, managing the cultural landscape on a larger scale and overall development of the local community in a sustainable direction. So close cooperation with the local community must continue in order to maintain local and regional involvement and an understanding of the need to protect the outstanding universal value. Imparting information is vital for this work, and new media, including social media, are increasingly being used.

Another presupposition is that the World Heritage site continues to be prioritized in all national heritage policies which pave the way for local value creation through projects that, at the same time, safeguard the natural and cultural heritage. New initiatives are particularly important. Since its establishment, the Vega Archipelago World Heritage Foundation has been cooperating with national authorities to develop Norwegian heritage policy. This work will be given priority in the years ahead. National support for World Heritage sites will be decisive to ensure a dynamic local community and give young families the chance to return to live on Vega. At the same time, it is important to work together with the rest of the region to develop sustainable tourism where all those involved help to safeguard and develop their assets.

Community ownership

Vega itself, as a local community, took the initiative to seek World Heritage status. Not all communities have this possibility. However, involvement and participation in the nomination process will be all the more important, especially in cultural landscapes, because the task of safeguarding and developing them is demanding. There are many conflicting interests and any development work must ensure balance and reciprocity between them.

Our experience is that local involvement, anchoring and cooperation around the same concept are the key to success. A good way to ensure this involvement may be through an organization that gives local stakeholders the opportunity to participate. Everyone must be able to feel that they are contributing and carrying out an important part of the job. At the same time, the national authorities must follow up through prioritizing the World Heritage site by obtaining knowledge, building expertise and giving access to funding for safeguarding and development.

To help people to work together, it may also be important to have a locally based World Heritage coordinator who is not part of the management authority, one who does not represent any special interests and can thus be seen as 'neutral' and independent.

Cultural landscapes exist in living local communities that are undergoing development. It is therefore very important to have a good Management Plan which ensures that all the players are agreed on a common vision of the future for the World Heritage site. In the Vega Archipelago, awareness of the eider tradition is deeply anchored, irrespective of the recent World Heritage status which indicated that it is also of significance for all humanity. This recognition is extremely valuable as a basis for our work. However, opinions may also differ as to how the outstanding universal value is to be sustained and which changes are acceptable. A description of a common vision of the future would therefore provide clarification.

2 Urbanism and Sustainable Heritage Development

The formation of urban centres has historically been a particular focus of scholarship and heritage conservation. In recent decades the processes of urbanization have become accelerated and heritage centres are taking on the full brunt of the impacts of diverse forms of globalization. At the same time as celebrating the 40th Anniversary of the World Heritage Convention, we are crossing another critical threshold in the history of humanity, with more than half the population of the world living in cities and towns. The case studies in this chapter illustrate the range of community engagement approaches taken in different parts of the world to address the challenges of conservation and sustainable development at World Heritage sites.

At the Historic Bridgetown and its Garrison site (Barbados), the key challenge is for heritage practitioners to help Barbadians come to terms with the historical legacies of slavery and colonialism, while forging an 'independent' identity.

The implementation of projects in the Medina of Marrakesh (Morocco) during the past decade has had a positive impact on the living conditions of people within the historic city. These include restoration of the sewage system, paving of lanes, repair of public fountains, creation of small parks in various places, inventory of houses at risk of collapse and the revalorization of the old urban fabric.

In the town of Luang Prabang (Lao People's Democratic Republic), constituent villages to benefit from infrastructure projects were selected on the basis of the inhabitants' willingness to participate in a 'village contract' for the maintenance of the streets, streetlights and sewage systems. The neighbourhood committees which existed under the traditional system of the *phu-baan* (village headman) thus became instrumental in the development strategy based on the site's outstanding universal value.

In Hoi An Ancient Town (Viet Nam), a former chairman of Hoi An People's Committee once famously pronounced that without the World Heritage site, Hoi An would die. In the context of rapid growth, the development of Hoi An, conservation of the Ancient Town and the transformation of the management authority, the safeguarding of the site's values through sustainable development is building up relationships with local communities.

The local World Heritage site community in the Historic Centre (Old Town) of Tallinn (Estonia), is driven by respect for the internationally and domestically recognized values of a unique and authentic historic city.

The Historic Centre of Santa Cruz de Mompox (Colombia), with more than 500 years of layered history and heritage values virtually intact, together with an intangible heritage legacy that holds potential, has a firm relationship with the economic activities of its residents.

6

Heritage and communities in a small island developing state: Historic Bridgetown and its Garrison, Barbados

TARA INNISS[1]

A momentous occasion

Barbados received its first World Heritage inscription in June 2011 for Historic Bridgetown and its Garrison, an outstanding example of British colonial architecture consisting of a well-preserved old town and a nearby military garrison.[2] It was a momentous occasion for the small island developing state (SIDS), which although a long and active State Party to the World Heritage Convention, continues to experience some challenges in ensuring the protection and preservation of the island's colonial and post-colonial cultural heritage. At the community level, heritage development has been stymied by several factors, but the central problem remains one of 'identity'. How do heritage practitioners help Barbadians come to terms with the historical legacies of slavery and colonialism, while forging an 'independent' identity?

Heritage practitioners are drawn from various organizations and community groups, representing diverse interests. The Management Plan for Historic Bridgetown and its Garrison must engage the active participation of all its stakeholders, but especially those situated in the country's urban and rural communities, in order to make the inscription meaningful and uphold World Heritage ideals. This case study suggests how property managers can mobilize community-based support for heritage in SIDS given the peculiar challenges of representing the island's past.

Located on the sheltered south-west coast of the island state, Historic Bridgetown and its Garrison developed as a colonial port town and entrepôt

[1] Department of History and Philosophy, Cave Hill Campus, University of the West Indies, Barbados.
[2] The 187-ha property was inscribed under criteria (ii), (iii) and (iv).

69

The cosmopolitan culture of Bridgetown, drawn largely from Africa and Europe, produced outstanding examples of architecture.

for the trans-shipment of goods and services in the lucrative sugar trade that dominated the British Atlantic economy from the 17th to 19th centuries. The Statement of Outstanding Universal Value notes its distinctive 17th-century organic street layout resembling English market towns, which is not found in any other Caribbean territory or in the British Americas. The town's fortified port spaces remain inextricably linked along the Bay Street corridor from the historic town centre to St Ann's Garrison, circling around Carlisle Bay, which provided safe harbour for the many military and commercial vessels that made Bridgetown the first port of call after the gruelling trans-Atlantic journey. The property was inscribed because of its retention and evolution of administrative, commercial, cultural and residential functions within the colonial and post-colonial urban space. Its cosmopolitan culture, drawn largely from Africa and Europe, produced outstanding examples of architecture, including St Ann's Garrison, colonial warehouses and dock facilities, including the Bridgetown Dry or 'Screw' Dock (the only such facility remaining in the world).

Heritage at a crossroads

Heritage development in the English-speaking Caribbean has been challenged by several factors as many states have emerged from their colonial status over the past fifty years. Like its counterparts in the region, Barbados

Classic Caribbean Georgian Styling at St Mary's Church.

has had to confront the historical legacies of slavery and colonialism and its new identity as an independent country, which has often fuelled conflict and contest over perceptions of the colonial past and its representation in the present. In addition to the philosophical challenges of representing the past, Barbados has been thrust into a globalized economic structure where the much-sought-after heritage tourism market demands heritage products and experiences for foreign consumption; forcing its tourism-based economy to reconcile these challenges expediently to take advantage of changing trends in the tourism market. Furthermore, in SIDS such as Barbados, there is increasing pressure placed on overstretched budgets which have to achieve several social development goals in health and education as well as maintain the stability required to secure foreign investment and positive economic growth.

Heritage, therefore, is at a crossroads in the island state – should heritage be used as a plank for sustainable development and planning or should the market determine what is appropriate for tourism consumption? The inherent tension between heritage and economic development through the tourism model is not lost on the Barbadian populace, and most acutely, on the local communities within the property. Even before the economic gains of tourism can be realized, Barbadians at the individual and community level must be able to confront their past on their own terms to participate meaningfully in the decision-making and effective management of the property.

Management of the World Heritage property

The Barbados World Heritage Committee, chaired by the Chief Town Planner (Town and Country Planning Department), is a multisectoral

The inherent tension between heritage and economic development through the tourism model is not lost on the Barbadian populace, and most acutely, on the local communities within the property.

Cabinet-appointed committee made up of technical experts from government-based and non-governmental agencies who oversee the management of World Heritage in Barbados, including Historic Bridgetown and its Garrison. As a SIDS, the Government of Barbados recognizes that in a small centrally planned economy, it was necessary to integrate World Heritage management principles in the daily operations and long-term planning of government portfolios. The Management Plan for Historic Bridgetown and its Garrison integrates existing policies and programmes, such as the Physical Development Plan (PDP, amended 2003), to guide the effective sustainable management of the property. Within the Management Plan, there are several Action Plans, including the Action Plan for Public Education, which has proven to be the most effective tool in raising awareness about the World Heritage property at both the community and personal levels.

Working with communities: Action Plan for Public Education

Since emancipation in 1834, working-class Barbadians have distanced themselves from their plantation slavery past, which set them on a deliberate quest for the next 150 years to attain social justice in the development of an independent state built on social democratic principles. Since independence in 1966, Barbados has achieved a high degree of human development. In 2011, it ranked 47th in the 2011 United Nations Human Development Index and it was the third highest-ranking country in the Latin American and Caribbean region, and was first in the Caribbean subregion). The tremendous gains Barbados has achieved in its social development, offering Barbadians access to free primary and secondary education and a free healthcare system, have transformed a society that once had one of the most coercive labour regimes in the region with few opportunities for education, as well as poor health standards.

In its post-colonial incarnation, no wonder that there was an erasure of a past built on slavery and colonial rule, which is one of the reasons why any of the tangible edifices built during the colonial period have remained neglected. Advocacy for the maintenance and preservation of buildings and spaces has long been a preoccupation of a small resident and ex-patriot elite who are perceived to be nostalgic about a past that many others wish to forget. The majority of African-descended Barbadians simply feel alienated from buildings and spaces that have become symbols of their colonization and enslavement.

Challenging this perception has been the first major obstacle for site managers. The Action Plan for Public Education has been the main vehicle for ensur-

The Main Guard (with clock tower) now serves as a focal point for activities around the garrison.

ing that Barbadians understand their role in enhancing and protecting the property's outstanding universal value. In fact, it can be argued that the Action Plan for Public Education undergirds all aspects of the Management Plan (i.e. Action Plans for Protecting, Preserving and Enhancing Heritage; Traffic Management; Tourism Management; etc.) which is overseen by the Barbados World Heritage Committee.[3]

Since inscription, stakeholders in the delivery of the Action Plan for Public Education have initiated mainstream approaches to public awareness, such as a Public Relations campaign using media spotlights, advertorials, broadcasts and a Facebook page. As an immediate response to the World Heritage designation, a local publisher of tourism-related market material, Miller Publishing, dedicated an entire issue of the *Ins and Outs* magazine to honouring the historic achievement. The magazine is distributed widely, mostly in the hospitality industry, but it has also become a popular keepsake for Barbadians.

However, site managers also recognize that this approach has a limited impact if not followed up with an intensive sensitization programme that occurs at all levels: individual and community outreach, as well as throughout the private and public sectors. Sensitization programmes are under way using a model developed by the Barbados Museum and Historical Society and the University of the West Indies (UWI) through a series of sessions called Barbados World Heritage Working Groups. Technical experts drawn from the Barbados World Heritage Committee are invited to speak with small groups of stakeholders to fulfill the immediate need for outreach, guidance and action among them. The sessions also encourage stakeholders to create a working relationship with the Barbados World Heritage Committee.

They have been very successful, but only reach small target groups which are responsible for delivering Action Plans, and wider outreach must also be considered using the media through engaging documentaries and events. However, with limited financial and human resources (inherent challenges in any SIDS, especially in the current economic crisis) to engage Barbadians,

73

[3] Government of Barbados, 2011, Management Plan for Historic Bridgetown and its Garrison.

West wing of Parliament and Clock Tower.

By acknowledging that their ancestors played a critical role as craftspeople in the building of Historic Bridgetown and its Garrison, many participants are no longer ambivalent about their role in the protection and enhancement of the property's heritage values.

progress has been slow and monitoring and evaluation of their effectiveness will need to be proven over time.

Based on feedback from various public education sessions, the Ministry of Culture has been hosting free public tours of the property commemorating aspects of national history, such as the National Heroes. These have been overwhelmingly popular among Barbadians wishing to learn more about their own history through the designation. Coupled with messages being transmitted through the media and various other public events, such as panel discussions and consultations, the World Heritage message is penetrating. Most interestingly, feedback from such activities suggests that since the designation and the increased number of public awareness initiatives, Barbadians are beginning to see themselves in the development of their own heritage. By acknowledging that their ancestors played a critical role as craftspeople in the building of Historic Bridgetown and its Garrison, many participants are no longer ambivalent about their role in the protection and enhancement of the property's heritage values.

Working with teachers

One of the first targets for sensitization programmes has been a series of workshops hosted by the Ministries of Culture and Education for primary and secondary teachers who teach history and social studies from across the island. Historians from UWI presented various aspects of the property's outstanding universal value. The workshops were designed to introduce teachers who deliver the social studies syllabus (primary and secondary levels) and history (at secondary level) to the property's values and develop teaching strategies and tools to integrate its historical development in the existing syllabus, which already explores some content relating to culture, history and heritage.[4] The following is an analysis of teachers' written reflections about the workshop:

- Deeper analysis and discussion about the process of African enslavement in the urban context to reveal the contributions of African-descended persons to the tangible and intangible heritage of the property.
- Discussion of the complex post-emancipation continuities and transformations that shaped the cultural and political development of the property so that the past is inextricably linked in narrative to the present.
- Connections must be made between the tangible (built) heritage of the property and its vibrant intangible heritage (i.e. marketing; music; street performance; Landship;[5] masquerade; folktales; etc.).
- Examination of the possible economic opportunities and benefits of World Heritage designation for participants; and,
- Expert exploration of the property (i.e. a field trip) with a knowledgeable guide who can reveal the stories of African-descended Barbadians as embedded in places and objects that demonstrate their political and cultural contributions to life within the property.

'I have passed by these historic locations several times in the past and been unaware of their history. I have developed more of an appreciation for our Barbadian heritage and a determination to be part of its conservation and preservation.'

History came alive during the field trip, as many had never fully appreciated the historic significance of these spaces before the workshop and their relevance to the development of Barbadian cultural heritage. Responses included:

- 'The opportunity to learn about the World Heritage aspect of Barbados is one that I will long remember. Our heritage is awesome and it is sad that it has been taken for granted and ignored.'
- 'I have passed by these historic locations several times in the past and been unaware of their history. I have developed more of an appreciation for our Barbadian heritage and a determination to be part of its conservation and preservation.'[6]

75

---

[4] Participants were invited to a two-day workshop. Day 1 was dedicated to class-based presentations on World Heritage and the outstanding universal value of the political, commercial, military and cultural development of the property. Day 2 comprised a three-hour field trip conducted by a UWI historian and a two-hour workshop to develop teaching strategies and tools based on the sessions and the existing syllabus.

[5] The Landship is a unique Barbadian tradition in which members masquerading in colourful navy uniforms dance on land as if on ship moving to the rhythm of the sea in 'manoeuvres' with a Tuk Band. The ceremony was associated with Friendly Societies which were organized to assist their members through hard times with funeral and sickness benefits.

[6] Extracted from anonymous written reflections composed by participants in the World Heritage in the Classroom workshops hosted by the Ministry of Education and Human Resource Development and the Ministry of Family, Culture, Sport and Youth, February 2012. Reflections were submitted to the UWI Department of History and Philosophy for monitoring and evaluation purposes.

Teachers who teach history and social studies from across the island are introduced to the property's outstanding universal value.

76

Extensive sensitization is required for the effective management of the property so that all stakeholders are engaged, not only to receive these messages, but to transmit them as well.[7]

Local community engagement with the property

Established commercial owner-operators within the property will be targeted by existing outreach initiatives. The residential communities span several levels of socio-economic development, including some more affluent precincts and several economically depressed areas. As a start, one group representing some of the residents and commercial interests in the Garrison Historic Area has organized itself into the Garrison Consortium Inc. The economically depressed areas of Nelson Street (near the southern end of Bridgetown on Bay Street) and neighbourhoods around St Mary's Church (near the western end of Bridgetown) are home to communities that require attention, not only to address some critical social development issues, but also to uncover their heritage potential.

Strategic participatory community-based initiatives are required for the sustainable development of the property, and especially in communities that have

[7] There may be scope for members of the Barbados World Heritage Committee to access funding through the Arts and Promotion Fund to produce short documentaries and arts programmes highlighting the World Heritage designation.

Codrington College, one of the oldest educational institutions for the propagation of the Gospel in the West Indies.

been traditionally overlooked. It is anticipated that the overall Management Plan will adopt models that have been used with success in other economically depressed World Heritage communities. It will also seek innovative ways to stimulate sustainable urban development through heritage by coordinating several agencies such as the Barbados World Heritage Committee, NGOs, the Ministry of Housing and Lands, the Ministry of Social Care and the Urban Development Commission.[8]

Site managers need to address the social development issues facing each community to unlock the heritage potential of sustainable community-based initiatives that can support poverty alleviation. At the moment, there has been limited coordinated outreach in these areas, but a sustainable cultural heritage programme based on research and social development designed in consultation with the people who live and work within these communities can be used as a strategy to support their economic and social development.

A first step in the research needed to support a sustainable cultural heritage programme is to evaluate the historical development of communities such as the Nelson Street Area, which developed along the Bay Street corridor that linked Historic Bridgetown to its Garrison. Its current formation dates to the mid to late 19th century, and its housing stock has a good mixture of iconic Barbadian commercial-residential properties and chattel houses, as well as some remaining suburban villas located on the River Road boundary. Historically, the area

[8] Sustainable heritage development models may be found in other SIDS World Heritage properties such as the Historic Town of Vigan (Philippines) and Aapravasi Ghat (Mauritius).

A Zoave militia man lowering the national flag during the changing of the guard.

developed in response to the rapid urbanization that took place after emancipation and the continuation of a bustling trans-shipment point for trans-Atlantic goods distribution to neighbouring islands in the schooner trade.[9] A survey of late 19th-century burial records for nearby Anglican St Mary's Church (located in Cheapside in the City) and St Paul's Church (located on Bay Street, which serviced the Garrison), reflects the occupational profile of the area known as Wellington Street, Nelson Street and Rebbit's Land. The men and women who populated the area in the late 19th century were skilled artisans and tradespeople (many of whom were Afro-Barbadians) who serviced the maritime industry and the mercantile orientation of the town.[10] In its 21st-century development, the area has become associated with crime and poverty, but in the late 19th and early 20th centuries it was home to an industrious and skilled workforce that supported the shipping activities of the town, leaving an indelible mark on its cultural heritage.[11]

[9] Bridgetown's settlement area expanded beyond the historic 17th-century centre of Cheapside and Broad Street (located northwest of Constitution River) with a growing population attracted to the employment and migration opportunities in the town's maritime-mercantile economy.

[10] Residents included shopkeepers, turners, mechanics, domestic labourers, porters, stevedores, laundresses, seamstresses, hucksters (vendors), needlewomen, carpenters, joiners, tailors, teachers, and most notably, seamen and master mariners. In the late 19th and early 20th centuries, residents chose these communities because they were near to the maritime-mercantile activities of the port and availed themselves of economic opportunities such as freighting goods and offering laundry services to ships anchored in Carlisle Bay. Similarly, oral history has revealed that some of the residents of these communities today seem to be third- and fourth-generation who have been continuously employed in port activities – today they work in the Deep Water Harbour.

[11] Extracted from the burial record database www.tombstones.bb for St Mary's Anglican Church and St Paul's Anglican Church (1860–1950).

Citizens walk along the shore of the historic harbour of Carlisle Bay, vital for the British Caribbean in the late 18th and 19th centuries.

The Pinelands Creative Workshop (PCW), a Barbadian NGO, has been active in community development initiatives in the Nelson Street area since 2000. In 2004–2006, the PCW's Sustainable Development in Low-Income Communities (SDLIC) programme received funding for a US$100,000+ project (partially funded by UNESCO) to use public education and the arts to develop a social consciousness in the community as well as to facilitate economic and social opportunities for young persons to build sustainable livelihoods. The project initiated several interventions to address crime, unemployment and training deficits after consultation with members of the community. It was revealed that there was a demand for training, entrepreneurial development and notably, for the fine and performing arts.[12]

Next steps

The recommendations for future development, especially in public education and training, are instructive and correspond to similar approaches that were taken to address teachers. The public education component needed to 'present learning in a creative and participatory manner'. Moreover, PCW's sustainable development programme determined that successful interventions required an informed but participatory approach that allows participants to determine

[12] Pinelands Creative Workshop. "Case Study: Sustainable Development in Low-Income Communities, Nelson Street, Bridgetown, Barbados", submitted to UNESCO (2006).

Aerial view of the National Heroes Square with radiating serpentine streets. The distinctive 17th-century organic street layout resembling English market towns is not found in any other Caribbean territory.

their own needs to identify programming interventions. Research was also cited as a critical part of the success of any programme within the area.[13]

The Barbados World Heritage Committee and other government or NGO partners can build on the knowledge derived from social development programming that has already taken place in the Nelson Street area prior to World Heritage inscription. It can recommend and initiate projects through its partners to use the arts and heritage to build peace within these communities and promote social consciousness and economic development.[14]

The Nelson Street area, as well as several low-income communities within the property, has its own role to play in the sustainable enhancement and maintenance of Historic Bridgetown and its Garrison's outstanding universal value. It has its own set of cultural heritage attributes that make its development within the context of Historic Bridgetown and its Garrison significant, and if future interventions could leverage its central placement within the property, then considerable gains can be made both for the community and for World Heritage.

[13] In its 2006 report, PCW noted that residents wished to see the development of a performing arts programme as well as sports development and an environmental programme. Pinelands Creative Workshop, Case Study: Sustainable Development in Low-Income Communities, Nelson Street, Bridgetown, Barbados (submitted to UNESCO, 2006). http://pinelandscreative.org/pcw/index.php?option=com_content&task=view&id=32&Itemid=170

[14] D. Thorsby,. 2008, *Culture in Sustainable Development: Insights for the Future Implementation of Art. 13 [Convention on the Protection and Promotion of the Diversity of Cultural Expressions]*, http://unesdoc.unesco.org/images/0015/001572/157287e.pdf; C. Aas et al., 2005, Stakeholder collaboration and heritage management, *Annals of Tourism Research*, Vol. 32, No. 1, pp. 28–48. Existing and proposed Barbadian legislation (Tourism Development Act and the Cultural Industries Bill) can help to provide economic incentives for community-based cultural heritage development if utilized appropriately.

Sustainable urban development

One of the Urban Development Commission directors recently emphasized that urban development is sustainable if the housing stock is improved and resources are distributed in entrepreneurship schemes with transparency and accountability. Issues of social cohesion, peace-building and environmental management are critical.[15] Historic Bridgetown and its Garrison is a lived-in city, and the Justification for Inscription demonstrated that the outstanding universal value of the property's built heritage and historic spatial organization has been shaped for almost 400 years by the people who lived and worked in the town. Building on the work that has already been done using arts and creative enterprise, as well as the knowledge gained from recent World Heritage public education initiatives, the Barbados World Heritage Committee can help to strengthen urban revitalization with heritage.

In order to safeguard the outstanding universal value the Committee wishes to develop a closer working relationship with all the communities within the property, especially those that can gain most from the inscription. Such communities must accrue the economic benefits of World Heritage as much as any other stakeholder, and in fact, if Barbadians want to enhance and protect the values of the property, an intensive outreach programme that addresses the opportunities and benefits of World Heritage as a tool for poverty alleviation needs to begin in communities. Moreover, to forge social cohesion in all the communities, an active and inclusive approach must be taken to promote cultural heritage to strengthen identity, build peace and promote environmental sustainability.

[15] *Sunday Sun* (national newspaper), 2012, Urban Rescue Mission: Interview with Derek Alleyne, Urban Development Commission Director, 4 March, A22.

81

7

The Red City: Medina of Marrakesh, Morocco

AHMED SKOUNTI[1]

A medina among medinas

Marrakesh, known as the 'Red City', is the largest of the thirty-one historic living towns (medinas) in Morocco with an intramural surface of 640 ha (including the Aguedal and Ménara gardens), extensive ramparts and their majestic gates, numerous monuments and residences, preserved gardens, long-inhabited markets and a vibrant craft industry. The cultural space of Jamaâ El Fna square mediates between the Medina and the external world. As an attractive interface and place of integration of populations originating from diverse backgrounds, it adds heritage value to the special role played by the Medina and the whole of this urban area in Morocco (Bigio, 2010). The population of the Medina accounts for 17.17 per cent[2] of that of the urban area of Marrakesh, i.e. 182,637 of 1,063,415 inhabitants, according to the 2004 census, and it represents a quarter of the population of the old cities of Morocco, i.e. 182,637 of 737,945 inhabitants (Taamouti et al., 2008).

Marrakesh was born out of strategic necessity. It was founded by the Almoravid dynasty in AD 1070–1071 on what seems to have been a space of commercial exchanges between mountain and plains communities. It was quasi-sacred territory, Amur, in which violence was banished, under the protection of a Berber divinity, Akuch. The sacred space of Akuch, or more precisely Amur Akuch, became Marrakesh, thus giving its name to the early urban settlement (Toufiq, 1988; Skounti, 2004). It was the historical capital of North

[1] Anthropologist, National Institute of Archaeology and Heritage Sciences (INSAP).
[2] This has decreased since the 1994 census (28 per cent). See the website of the Haut Commissariat au Plan: www.hcp.ma.

Africa and one of the important cities of the western Mediterranean basin in the Middle Ages. The monuments resulting from various periods testify to the tumultuous history of the rise and fall of Marrakesh.

The Medina of Marrakesh was inscribed on the World Heritage List in 1985 and the following four criteria illustrate its outstanding universal value:

(i)   Recognition of the impressive number of masterpieces sheltered by the Medina in the fields of architecture and art, each one of which could justify recognition of outstanding universal value;

(ii) Acknowledgement of the urban qualities of a historical capital having exerted a decisive influence on later urban development, in particular on Fez;

(iv) Consideration that Marrakesh, which gave its name to the empire of Morocco, is a completed example of a major Islamic capital of the western Mediterranean; and

(v)   Highlights a historic living city rendered vulnerable due to demographic change.

The inscribed property consists of two entities: the Medina itself with its southern prolongation consisting of the Agdal gardens and the Menara olive groves, the 13th-century basin and the 19th-century pavilion. These constitute the historical heart of the urban area of Marrakesh also known for its 1,000-year-old palm grove (Palmeraie) with a significant number of date palm trees. The Medina is inhabited (population 182,637) while the Menara has a protected historic building in the centre. Among the eight Moroccan World

83

Restoration of Badii Palace in 2011.

Real estate speculation caused heated debate among scholars, some of whom felt that it was laden with the threat of neo-colonialism.

Heritage sites, Marrakesh is the only one to be listed for its impressive number of masterpieces of architecture and art under criterion (i).

Challenges and transformations

At the beginning of the nomination process the communities played a limited role as it was driven by the Ministry of Interior through its Department for Urbanism. World Heritage status had little impact on the population until 2000. However, the Ministry of Culture in its local service for conservation was able to defend the site from real estate appetites and urban development programmes. The development of tourism in the past decade and the settling of migrants who bought *riads*[3] and houses in the Medina posed challenges. In addition, a balance had to be struck between limited natural resources (mainly water) and cultural resources and the growth in population and tourism. A tighter relationship had to be developed between the site and the local and wider community. A project-driven approach is now leading to transformations in the conservation of the site.

The implementation of projects in the Medina during the past decade has had a positive impact on the living conditions of people within the historic

[3] A *riad* is a traditional urban residence organized around a central non-covered courtyard planted with trees, mainly orange trees.

In the heart of the Medina, Jamaâ El Fna square is inscribed under both the World Heritage Convention and the Intangible Heritage Convention. Safeguarding measures are being implemented together with the bearers of intangible cultural heritage.

The implementation of projects in the Medina during the past decade has had a positive impact on the living conditions of people within the historic city. These include restoration of the sewage system, paving of lanes, repair of public fountains, creation of small parks in various places, inventory of houses at risk of collapse and the revalorization of the old urban fabric.

city. These include restoration of the sewage system, paving of lanes, repair of public fountains, creation of small parks in various places, inventory of houses at risk of collapse and the revalorization of the old urban fabric. The gentrification of the Medina from the end of the 1990s gradually raised its image. However, the prices of the houses and the *riads* sky-rocketed. The real estate speculation allured many foreign buyers, mainly Europeans, to Marrakesh. This phenomenon caused divergent opinions among scholars. Some of them considered it as a complex phenomenon and an opportunity for the safeguarding of houses otherwise threatened with decay and collapse (Kurzac-Souali, 2006; Skounti, 2004). Others felt that the real estate growth is laden with the threat of neo-colonialism (Escher et al., 1999; Escher, 2000). The complexity of this phenomenon lay beyond these two points of view, in the interaction between the local communities and the immigrants (Saïgh Bousta, 2004). The local authorities themselves were quickly overwhelmed and found themselves not equipped to manage the impacts of rapid growth. Despite the initial lack of capacity to mitigate negative impacts on the World Heritage site, a new law was promulgated on 18 December 2003 to preserve the guest houses that now occupy hundreds of *riads*.

85

Architectural Charter of the Medina of Marrakesh

The restoration and rehabilitation of the old houses by new owners with different aspirations and cultural backgrounds brought about new challenges in the transformations of the *riads*. The relative respect of height was the common ground of their renovation, rather than restoration projects. The Medina Charter was finally adopted in 2008 by the Urban Agency of Marrakesh in cooperation with the Regional Inspection of Historic Monuments and Sites.

The Medina Charter adopted in 2008 established a typology of the architectural and urban characteristics helping to improve the relationships of public authorities and private individuals. Here, patio of the Ben Youssef *madrasa*.

This document established a typology of the architectural and urban characteristics of the historic city by recognizing a hierarchy of spaces, neighbourhoods and architectural elements. The Charter was followed up with a set of regulations that made it possible to improve the relationships of public authorities and private individuals within the traditional urban fabric and its conservation. The Charter dealt with private properties, public buildings, commercial and service spaces (including guest houses), façades, infrastructure for drinking water, the street signage and urban planning. Exceptional cases which require in-depth impact studies concern certain types of facilities such as swimming pools, elevators and basements of houses. Building construction and modifications exceeding 8.5 m in the Medina and change of purpose of buildings from their original use or adaptive re-use had to be approved. The Charter is used by the administration in charge of heritage and town planning to control building and construction work within the World Heritage site.

The conservation efforts are coupled with the gradual decrease in the local population of the Medina and this appears to have contributed to improving the living conditions inside the World Heritage property. The effect can be measured in the coming years. Even if this depopulation is slow (about 7,000 inhabitants in less than a decade, 1994–2004), it seems to be irreversible. An index of this impact is the alarming practice of parcelling out of land because of distribution of the houses between heirs or the renting of rooms of the same residence to several different tenants. Lastly, major social changes in the Moroccan family, in particular the transition from an extended family to a nuclear one, are unquestionably reflected in the occupancy rate of built space which had reached a critical point in the new millennium.

Tourism is another significant factor to be analysed when dealing with sustainable development in the Medina of Marrakesh (Tebbaa, 2010). There is no in-depth survey on the branding role of World Heritage in the attractiveness of the city but it obviously contributes to it, along with the intangible cultural heritage element of Jamaâ El Fna square. Marrakesh accommodates approximately 1.5 million tourists a year for a population of approximately 1 million inhabitants. The city has 130 classified hotels and several of them are located in the Medina, along with 578 guesthouses.[4] Tourism in Marrakesh is a combination of mass tourism, luxury tourism and convention business. The average duration of stay is four days. Water consumption quickly became a serious concern in a rather arid region where annual rainfall does not exceed 300 ml. The sustainability of the natural resources, in particular of water (El Faïz, 2002), was the focus of recent debates on the most suitable model for responsible tourism. Vision 2020 for tourism in Morocco, a recently adopted national strategy for this sector, made sustainability one of its pillars.[5]

Marrakesh's Koutoubia Mosque, in an advanced state of degradation in recent years, urgently needs a visitor management strategy.

87

One of the measures for the sustainability of water resources is recently addressed through the construction of the sewage treatment plant of Marrakesh. Inaugurated on 29 December 2011, it covers a surface area of 17 ha with a budget of 1.23 billion dirham. It is a strategic partnership between the Autonomous Agency of Water Supply and Electricity of Marrakesh (RADEEMA), the state and the tourism industry. From the 43 million megalitres of waste water generated by the city each year (including water from the World Heritage site), 33 million megalitres of recycled water is produced by the plant. It will be used for the irrigation of green spaces and golf courses. In parallel, an extensive programme of conservation

[4] See the web site of the Association des Maisons d'hôtes de Marrakech et du Sud: www.amhms.com.

[5] Vision 2020 for tourism was adopted by the Moroccan Government under the patronage of King Mohammed VI on 30 November 2010 in Marrakesh. See: www.tourisme.gov.ma.

Frieze detail at the Ben Youssef *madrasa*. Of the eight Moroccan World Heritage sites, Marrakesh is the only one to be listed for its impressive number of masterpieces of architecture and art.

and development of the Marrakesh palm grove was launched in 2007 by the Mohammed VI Foundation for Environmental Protection. It focuses on education and publicity campaigns, a project on a museum of the oases and the plantation of hundreds of thousands of palm seedlings (by 2 January 2012, 456,160 seedlings had been planted).[6]

Sustainability strategies also address the continuity of cultural resources. The Periodic Report of 2010 on the state of conservation of the World Heritage property showed that the Medina is exposed to negative threats such as the uncontrolled development of trade, infrastructure of services and transport. It also pointed out factors induced by the gentrification such as threats to social cohesion, changes in population characteristics, erosion of traditional lifestyles, disappearance of the traditional knowledge of management and proliferation of tourism-related activities. The report underlined the threats to visual integrity, the aesthetic aspects of façades, vibrations and pollution from motorized transport, urban development around the Medina which is likely to affect its visual perimeter and the disappearance of the system of underground water drains (*khettaras*).

Efforts are also being made to minimize impacts on the most visited cultural sites such as the Badii Palace, Bahia Palace, Dar Si Saïd Museum, Saadis Tombs, Medersa Ben Youssef and Ménara gardens, among others. Some of these have fallen into an advanced state of degradation in recent years, such as

[6] See the website of the Foundation: www.fm6e.org.

Ménara Gardens is one of the most visited cultural sites of Marrakesh.

the Almoravid Qoubba (Koutoubia Mosque). For all these sites of great heritage value, there is an urgent need for a visitor management strategy with an established ceiling on visitors and measurable key indicators of conservation. Most of the designated sites are the same ones as during the French Protectorate (1912–56). Many other sites could be prepared for visitors, such as the Dar El Bacha Palace, the historic gates of the rampart, the 13th-century bridge on the Tensift River, the Medersa Ben Saleh and the ruins of the Almoravid Palace, excavated in 1996 near Koutoubia Mosque, among others to be identified and assessed for their significance and contribution to the site's outstanding universal value.

Public-private partnerships

In all these projects the role of the local communities could be made more visible. In the past decade the real estate ownership by foreigners within the Medina has led to prices of houses and the *riads* reaching an unprecedented level. The real estate agencies proliferated in the Medina and in the new town. Some added to their basic services restoration, decoration and furnishing. Investment in the Medina was accompanied by the Charter and regulations from public authorities at a legal level or on the adaptive re-use of the old urban fabric. But few benefits accrued to the local inhabitants and

Marrakesh accommodates approximately 1.5 million tourists a year. Here, visitors discover Saadi Royal Tombs.

Safeguarding measures for Jamaâ El Fna square are being implemented with the bearers of intangible cultural heritage, recently organized in associations and aspiring to be full partners in any project of safeguarding their knowledge and skills.

small business owners who are very often obliged to leave their houses to live in apartments and district-dormitories without green spaces and socio-cultural infrastructure.

However, there are actions undertaken jointly by the public authorities and some categories of residents to alleviate poverty and to improve the household incomes and more generally the living conditions. Within the framework of the programme of the National Initiative for Human Development (INDH) launched by King Mohammed VI in 2005, the restoration and rehabilitation of eleven caravanserais within the Medina was part of this effort. Identified on the basis of a study launched in 2006, they were restored. Scenarios of rehabilitation were proposed, for example the attribution of new socio-economic functions such as craft workshops. Not only does the heritage status save them from abandonment and degradation but it gives them a second breath with positive impacts on the surrounding district. Another example is that of Jamaâ El Fna square, whose safeguarding measures are being implemented with the bearers of intangible cultural heritage, recently organized in associations and aspiring to be full partners in any project of safeguarding their knowledge and skills. Intergenerational transmission activities with a focus on young people are accompanied by internships, social support measures, healthcare and so on.

The public-private partnership within the Medina of Marrakesh is varied. Some projects led within this framework succeeded. First appears the

Erosion of traditional lifestyles and proliferation of tourism-related activities is being continually pointed out in recent years. Here, tanners in the Debbaghine district of Marrakesh, north-west of the old city.

Communal Development Plan (PCD) initiated by the Mayoralty very recently. The PCD is meant to cover the period 2011–2016. It was elaborated on the basis of a participative approach including the Mayoralty members, the civil society representatives, the university researchers and the private sector investors. The PCD comprises a series of projects selected by participants during workshops organized in 2010. These projects are dedicated to issues such as: basic infrastructure and urban circulation, fight against unhealthy habitat, access to basic services and fight against exclusion and poverty, urbanism and town planning and heritage, urban environment and sustainable development, local governance, social and cultural and sportive action towards civil society. The plan is now under implementation.

Ongoing World Heritage projects implemented by the Mayoralty with the contribution of its partners are shown in Table 1.

Other partnerships may also be regarded as constructive. First appears the involvement of the banks and the telephony companies as partners in the conservation of cultural heritage. The Banques Populaires Foundation financed the restoration of the three historical fountains of Bahia, Bab Aylen and Bab El Khmis. The intramural garden of Arset Moulay Abdeslam (17th century; 9.2 ha) was restored and rehabilitated into a cyber-park within the framework of a partnership between the Mohammed VI Foundation for Environmental Protection, the City of Marrakesh, the Prefecture and the company Maroc Telecom.

Table 1 Public-private partnership projects

Project	Location	Mayoralty	Private and public partners	Total (MAD)*
Urban infrastructure, system of roads, car circulation and lighting	Inside and around the Medina, palm grove and the Ménara and Agdal gardens	33 820 000	54 600 000	88 420 000
Restoration and rehabilitation work	Ramparts of the Medina, Agdal gardens, Agdal Ba Hmad gardens	4 500 000	44 000 000	48 500 000
Construction of ditches against floods	Around the Medina	–	45 000 000	45 000 000
Recreational infrastructure	Ghabat Chabab park between Medina and Ménara gardens	6 000 000	6 000 000	12 000 000
Total (MAD)		44 320 000	149 600 000	193 920 000

*US$1 = 8 MAD (Moroccan dirham).
Source: Mayoralty official website: www.ville-marrakech.ma.

International cooperation also contributes to the ongoing efforts of safe-guarding the Medina. In 2006–2008, the RehabiMed project financed by the European Union within the framework of the programme Euromed Heritage funded the restoration of three houses in the Medina in order to sensitize the residents to the virtues of traditional building techniques and the valoriza-tion of their neighbourhoods thanks to responsible restoration to professional standards.

Improved community involvement

The Medina of Marrakesh has gone through two periodic reporting cycles in 2000 and 2009. A range of concerns expressed in 2000 continues to be addressed and by 2009 the above-mentioned projects had made a positive impact on the conservation of the World Heritage site.

There is increasing heritage awareness and it is reinforced among the range of stakeholders involved in the management of the property. A clear vision is

Many people consider the neighbourhood monuments as 'tourist places', and the majority of them have never entered these temples of heritage, nor have their children. This is one example of how the involvement in the management of a World Heritage site of local communities living in precarious conditions is a long-term action.

called for in order to make tourism more sustainable, focusing on both natural and cultural resources. Improved benefits are desired for the communities living within the World Heritage property. The cultural heritage of the Medina could become a genuine driver of sustainable urban development. The existing cultural heritage as well as the 'sleeping' heritage can be mobilized to improve the living standards of local people, creating socio-economic and socio-cultural infrastructures and jobs. The local communities could play a major role in identifying and upgrading other sites and increase thus income from tourists, currently about 11 million MAD per year (1 million euros).[7]

Parallel to the current political system of representation (based on Mayoralty members elected by majority vote), improved dialogue with stakeholder groups is enabling better participation of local communities in the management and development of the Medina. The sense of ownership between local residents and the heritage site needs to be improved. Many people consider the neighbourhood monuments as 'tourist places', and the majority of them have never entered these temples of heritage, nor have their children. This is one example of how the involvement in the management of a World Heritage site of local communities living in precarious conditions is a long-term action.

93

Integrated heritage development

The Medina of Marrakesh with Jamaâ El Fna square at its heart, inscribed under two UNESCO Conventions (World Heritage Convention of 1972 and Intangible Heritage Convention of 2003), provide an opportunity for each element to benefit from the other thus complementing the commitment to conservation and safeguarding of their outstanding universal value (Skounti, 2009, 2011). Marrakesh could become a laboratory for integration of tangible and intangible heritage with a mandate under two international legal instruments, better involving local communities in the safeguarding of both tangible and intangible heritage. The safeguarding measures for the square cannot be successfully implemented without the bearers of intangible cultural heritage. Recently organized into associations, they aspire to be full partners in any project of safeguarding their knowledge. Their main demands from local communities are recognition and social rights such as allowances, health care and appropriate facilities for intergenerational transmission of their knowledge to young people. The intangible heritage could thus be a suitable means of safeguarding the whole property and its values.

---

[7] Source: Inspection of Historic Monuments and Sites of Marrakesh.

8

Capacity-building for sustainable urban development: Town of Luang Prabang, Lao People's Democratic Republic

MINJA YANG[1]

Value of inscription

At the confluence of the great Mekong River and its tributary, the gentle Khan River, is the Town of Luang Prabang in the northern Lao province of Luang Prabang. Its inscription on the World Heritage List in 1995, in recognition of its outstanding universal value by the international community, represented, for both government and people, an affirmation of national identity and independence, a hallmark after nearly half a century of war.

The initial elation over the World Heritage status, celebrated with great pride by not only the inhabitants of Luang Prabang but by all citizens of Laos, has become seventeen years later a source of debate in the choice of the city's future. What exactly is the outstanding universal value of this World Heritage site, which the government, on behalf of its current and future generation of citizens, has undertaken to protect, conserve and enhance forever? In what way will the values-based urban development strategy differ from those that other cities of South-East Asia have experienced by design or by default? Moreover, how will this choice in the options of development impact the multi-ethnic population of Laos given the geopolitical dynamics of the region, in the state we now know as the Lao People's Democratic Republic?

[1] President, Raymond Lemaire International Centre for Conservation, KU Leuven (Belgium); former Deputy Director, UNESCO World Heritage Centre; former Director, UNESCO Cluster Office in New Delhi.

Morning alms to monks. The Town of Luang Prabang has serene Buddhist temples and monasteries in each of its fourteen urban villages.

95

Outstanding universal value and cultural pluralism

Options for development of Luang Prabang, both as a city with 60,000 residents and as a province with a population of 425,000, must be understood in the context of Laos, a small land-locked country, bordered by Myanmar (Burma) and China to the north-west, Viet Nam to the east, Cambodia to the south and Thailand to the west. Estimated at 6.5 million in 2011, the population consists of twelve ethnicities and three main groups: the Lao Loum (low-land people), the Lao Theung (midland people), and the Lao Soung (highland people). The few towns of Laos which existed in pre-1975 were also home to Vietnamese, Chinese and Thai but many of them left after independence in the early 1950s, or fled as refugees with the advent of the Communist Pathet Lao forces in 1975.

The ethnic puzzle of Laos has marked the social structure of Laos, where the primary divisions were, and still are, determined on ethno-geographical criteria rather than on the bases of social class or economic role, although this is beginning to change. The boundaries of Laos, resulting from historical expediency, have thus never marked the living space of a single integrated Lao society. The future of the country, with its powerful neighbours sharing larger proportions of the same ethnic groups, will be determined by the governance of cultural pluralism, which is intricately linked to the management of

According to legend,
the Buddha smiled
when he rested in
Luang Prabang for a
day, prophesying that
it would one day be
the site of a rich and
powerful capital city.

The future of the
country, with its
powerful neighbours
sharing larger pro-
portions of the same
ethnic groups, will
be determined by the
governance of cultural
pluralism, which is
intricately linked to the
management of the
country's biodiversity
on which the means
of livelihood of almost
half of the population
is based.

the country's biodiversity on which the means of livelihood of almost half of the population is based.

As a landlocked country, Laos is dependent on its powerful neighbours. But with the dredging of the Mekong River financed largely by China, river transport will soon change the economics of the country. Its rich mineral resources will feed the ever-expanding industries of China. Moreover, the Great Asian Highway connecting continental South-East Asia has improved road transport, particularly with the suppression of pockets of ethnic conflict and banditry that continued even after the end of the civil war. Security has brought about new dynamics of regional trade and economic prosperity. Equitable economic growth, cultural and political rights for all ethnic groups is needed to sustain peace and avoid renewed conflict rekindled by social contradictions with far-reaching consequences of a transborder nature. In this context nationalism uniquely based on the dominant culture, as propelled by the symbolism of the World Heritage Town of Luang Prabang, needs to be understood.

The Town of Luang Prabang, with its serene Buddhist temples and monasteries in each of the fourteen urban villages, reflects the culture of the lowland Lao engaged in irrigated rice cultivation. But the town has also been, historically and to the present day, a market for the Lao Theung and the Lao Soung population inhabiting the surrounding hills. With more than 50 per cent of the

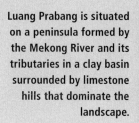

Luang Prabang is situated on a peninsula formed by the Mekong River and its tributaries in a clay basin surrounded by limestone hills that dominate the landscape.

97

total land area covered by forests, contributions by the minorities engaged in timber and non-timber production to the wealth of Luang Prabang Town must be duly recognized. Moreover, with the surge in economic growth witnessed over the past two decades deriving mainly from the phenomenal development of international tourism to Luang Prabang following its World Heritage inscription, it is all the more important that new opportunities are shared with all ethnic groups inhabiting the province.

Outstanding universal value as the guiding principle for sustainable development

The statement of significance attributed was 'Luang Prabang represents, to an exceptional extent, the successful fusion of the traditional architectural and urban structures and those of the European colonial rulers of the 19th and 20th centuries. Its unique townscape is remarkably well preserved, illustrating a key stage in the blending of two distinct cultural traditions.'[2] The outstanding universal value is based on the following:

■ strong nature-culture linkages shown by the integration of the built environment with the natural attributes of the river and the urban wetlands;

---

[2] The Town of Luang Prabang was inscribed on the World Heritage List on 4 December 1995, by the World Heritage Committee at its 19th session held in Berlin, based on criteria (ii), (iv) and (v).

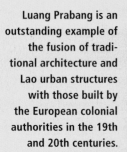

Luang Prabang is an outstanding example of the fusion of traditional architecture and Lao urban structures with those built by the European colonial authorities in the 19th and 20th centuries.

- harmonious co-existence of traditional Lao and French architecture, which can be seen in the building typologies;
- fusion, the blending of the town patterns of two different epochs and cultural traditions – the village pattern and the 'colonial' grid pattern.

Therefore to maintain the outstanding universal value of Luang Prabang, with its authenticity and integrity, it is vital that all future development in this World Heritage site protects the riverscape (view of and from the river) and the relationship between the built area and the river marked by the urban wetlands (ponds), urban and peri-urban agriculture; and the historic urban landscape featuring the hills, riverscape and the historic town morphology that shows the pre-colonial Lao villages inside the urban blocks formed during the French period, which experts note had initially been traced by the Siamese (now Thailand) shortly before Laos became a protectorate of France.

The buildings in traditional Lao style, including the historic temples, would need to maintain the original building materials and replace them only when necessary. Because of continuing traditions of craftsmanship and building technology, new Lao buildings will not necessarily undermine the site's authenticity. However, new construction of French colonial-style buildings, which is specific to the late 19th and early 20th centuries. could be considered pastiche under

Textile dyeing ranks among the many skills of local communities.

the norms of heritage conservation. The disappearance of significant buildings of Vietnamese style and Chinese shop-houses that testify to the important historic role of these communities in the urban economy is sadly erasing the social history of the town.

In October 1995, just before the World Heritage inscription, a UNESCO expert mission was sent to Laos to gain a better understanding of the legal and administrative framework of the national and provincial authorities whose responsibility it would be to maintain the outstanding universal value of Luang Prabang. The International Council on Monuments and Sites (ICOMOS) evaluation mission which had taken place earlier that year had recommended 'referral' because the urban plan which had been developed less than a year before by UN Habitat experts was deemed to require more time to demonstrate its effectiveness. While ICOMOS was correct to doubt the enforcement capacity of the government, it was also clear that without international oversight and technical cooperation, giving another year of trial would not

Urban pathways have been restored with a view to improving drainage and circulation while preserving the urban ambiance.

enable the building of local capacities. The Town of Luang Prabang in 1995 was therefore under serious threat of incongruent growth as investors from neighbouring countries aspired to develop a new mass tourism destination in an unspoiled environment.

The UNESCO mission was therefore to identify urgent measures to strengthen national heritage protection capacities. The mission was led by Yves Dauge, Mayor of Chinon (France) and an experienced urban planner and political personality, assisted by Francis Engelmann, then Director of the Lao National School of Administration and Management (ENAG) to look into training needs. Within two months following this mission, city-to-city technical cooperation between Luang Prabang and Chinon was established to reassure the World Heritage Committee that capacity-building concerns were being addressed. Engaging in a multi-year commitment initiated with a catalytic grant from the World Heritage Fund, the 'decentralized cooperation programme' is now in its seventeenth year of existence. It generated numerous grant aid projects from the European Commission and the French

Keeping the river banks free from construction as part of the urban plan is key to ensuring local agriculture while preserving the landscape.

Development Agency (AFD) bringing in total project funds of over €29 million. These projects, from conservation and development planning to upgrading of roads and sewers while maintaining the historic urban pattern, waste management and rehabilitation of the town's wetlands, urban agriculture and hydro-system, historic monuments conservation and housing improvement linked to heritage protection, have resulted in enhancing the site's outstanding universal value while serving sustainable development objectives.

Appropriate governance and management

In promoting heritage-based development, the Lao-French project team started with a review and subsequent revision of national laws and regulations on cultural heritage protection, urban development and environmental laws. Rehabilitation projects were simultaneously launched to draw the attention of officials and inhabitants to conservation norms and practice. An inter-ministerial National Committee on Heritage was established, presided over initially by the Minister of Culture but later by the Deputy Prime Minister to ensure compliance by all ministries. At the local level, the Luang Prabang Provincial Committee on Heritage headed by the Vice-Governor of the Province was also set up to mainstream heritage in all public programmes. The Maison du Patrimoine (MdP) or Heritage House, a technical unit to oversee all public and private works, was established in 1996 as a service within the Provincial Government and staffed by the Lao-French project team initially financed from international project funds.

The Conservation and Enhancement Plan (PSMV: Plan de Sauvegarde et Mise en Valeur) was developed through a training mode involving the provincial authorities and young graduates of architecture and engineering schools in Vientiane. On-the-job training took place through extensive work on documentation to develop heritage inventories and evaluation of building requests. Preparation of thematic maps of the town where the only existing base map dated back to the French colonial period turned out to be a major challenge for the surveyors. Provisional building regulations based on the outstanding universal value of Luang Prabang were soon established, pending completion of the PSMV to revise the existing urban plan of 1994.

Villages to benefit from infrastructure projects were selected on the basis of the inhabitants' willingness to participate in a 'village contract' for the maintenance of the streets, streetlights and sewers. The neighbourhood committees which existed under the traditional system of the *phu-baan* (village headman) thus became instrumental in the values-based development strategy.

To protect the 'nature-culture link', the urban wetlands, agricultural land and the hill area were designated as no-construction zones. To understand the hydrological system and the seasonal dynamics of the Mekong and Khan rivers, projects were developed for funding by the European Commission and the French Fund for the Environment. Particularly at risk were the 180 urban wetlands or ponds located within the World Heritage property boundaries which were mostly still in use for the cultivation of fish and vegetables. Serving also to drain rainwater during the monsoon season, these wetlands have for centuries played a vital role in the urban ecosystem. Many were however polluted with wastewater flowing into them, creating unsanitary conditions which had led to their land-filling, thus depriving the local inhabitants of important sources of food for consumption or barter trade. A survey conducted in 1999 within the framework of the hydro-system project under the scientific coordination of the Institut de Milieu Aquatic et Corridor Fluviale (IMACOF) of Tours Polytechnic, noted that some 80 per cent of the town's population were dependent at least partially on food produced in these wetlands.

Sewage and drainage extensions taking place under several development aid and soft-loan projects, however, were inappropriately designed with excessive costs and without considering the maintenance issues, which combined with

Local *tuk-tuk* transport is given priority to help local employment while preserving the city centre from traffic congestion.

land-filling of the wetlands were resulting in the aggravation of sanitary conditions in certain areas. The Environment Unit of the MdP with IMACOF experts therefore developed an action plan for the rehabilitation of the urban hydro system, including the design and construction of alternative narrow V-based sewers running along the neighbourhood paths and a wastewater treatment system of simple technology installed in several community-managed urban wetlands and in schools and monasteries. In the absence of an efficient provincial service for wastewater treatment, it was essential to foresee a system that was both affordable and simple to be maintained by the users themselves. Combined with public awareness-raising and educational activities, many training sessions were organized to enhance understanding of the functions of the wetlands and on health hazards of silted water, such as malaria and dysentery.

Villages to benefit from infrastructure projects were selected on the basis of the inhabitants' willingness to participate in a 'village contract' for the maintenance of the streets, streetlights and sewers. The neighbourhood committees which existed under the traditional system of the *phu-baan* (village headman) thus became instrumental in the values-based development strategy, not only for the protection of the wetlands but also for the town pattern that signifies their historic function in the interaction of the paths, streets and roads. The village headman of each neighbourhood also served as the intermediary to ensure that the residents complied with building regulations. A Fund for Housing Rehabilitation was also set up under the AFD grant, for the benefit of private owners of heritage buildings, serving to promote building compliance in exchange for donations of traditional roof tiles and technical support by the MdP's Architecture Unit.

These schemes required huge efforts of consultations with local residents, national and provincial government officials to convince them why the roads could not be widened to accommodate more cars or why off-street parking could not be facilitated by changing the alignment, or why addition of floors to create more rooms would not be possible in order to maintain the overall townscape. The residents, by and large, respect the building codes which unfortunately has not been the case with some investors who tried to buy their way to build as they saw fit. When non-compliance became a potential danger, such as for large-scale land-filling of wetlands, or demolition and reconstruction of the central market or the golf course-related real estate development, a State of Conservation report was submitted to the World Heritage Committee to redress the situation. Michel Brodovitch, the chief scientific advisor appointed by UNESCO and involved in the programme for over fifteen years, worked closely with the MdP architects to find solutions to satisfy the investors while respecting conservation needs. The MdP database set up with support by the Tokyo Institute of Technology which records every building permit has allowed the monitoring of change, including information on non-compliance. This architectural database was recently upgraded with the addition of the Geographical Information System (GIS) under the Tokyo Tech project, now in its eighth year of operation. Thanks to major investment in training, the Lao staff of the MdP Database Unit are now capable of maintaining the system. These information and communication technology (ICT) tools have proven to be most useful, also for public information and awareness-raising carried out through the ICT Centre of the MdP, open to the local community as well as to tourists.

Coherent territorial development

As in any living city, the needs for housing, commerce and public utilities have grown over time; therefore the next step of the MdP was to develop a regional

Boua Kang Bung wetlands have been rehabilitated and serve as an interpretation centre and 'green lungs' within dense urban areas.

105

plan called the Scheme for Coherent Territorial Development (SCOT) with technical support from Chinon funded by a second AFD project. The objective is to plan for socio-economic and physical infrastructure that the World Heritage core area cannot accommodate but nonetheless is necessary to improve the quality of life of the growing number of people migrating to Luang Prabang in search of opportunities. SCOT thus complements the PSMV to ensure a harmonious transition between the core area, the buffer zone and the surrounding territory to avoid chaotic spatial development, so that the entire region along the banks of the two rivers would maintain its attractiveness, thus expanding the catchment area for investment and development. SCOT and its regulations was finally approved by the provincial and national authorities in 2011, thanks to the 2007 decision of the World Heritage Committee requesting the State Party to establish a buffer zone and to maintain the natural surroundings of the core area.

Lessons learned

The experience of Luang Prabang has been rich and instructive, pointing above all to the primordial importance of public awareness-raising and capacity-building activities in all projects. Another lesson learned is the advantage of having all urban infrastructure aid projects executed from the same locally based office, which in this case has been the Maison du Patrimoine. The commitment of Chinon has been crucial as the main international technical partner, working with UNESCO in project formulation, mobilization of funds and experts, and in building the competence and institutional memory of MdP. While other World Heritage sites may not have the fortune of the substantive external aid that Luang Prabang benefited from, any government can insist on training components in aid and investment projects.

Vision for the future

The next objective to diversify economic opportunities beyond tourism has already started with the establishment in 2002 of the National University of Luang Prabang, which was strongly recommended by UNESCO to build knowledge-based industries by taking advantage of the international reputation and heritage resources of Luang Prabang. With town extension to the north towards the planned new bridge across the Mekong and national route deviation away from the World Heritage core area, the local community and investors now have space to develop new activities outside the sensitive area.

To bring further territorial coherence beyond the core and buffer zones, the development of another territorial plan on a much larger scale was initiated with the view of developing a Biosphere Reserve under the UNESCO Man and the Biosphere (MAB) Programme in the Khan River Basin.

Efforts to respect the outstanding universal values of the World Heritage site of Luang Prabang have thus served to provide a new vision of urban and territorial development beyond the core area. The sustainable future of Luang Prabang will depend on the protection of its cultural and natural heritage, but above all on inclusive growth based on cultural and natural diversity and cultural pluralism anchored on education, the sharing of development opportunities and democratic governance for the benefit of all citizens of Laos in its rich multi-ethnicity.

9 World Heritage in poverty alleviation: Hoi An Ancient Town, Viet Nam

AMARESWAR GALLA[1]

Lived-in heritage

Viet Nam is one of the few developing countries with impressive achievements in poverty reduction. Diversification of the resource base for local communities, responsive infrastructure development and expanding choices for the poor have been critical in the *Doi Moi* economic reform and poverty eradication programme. Further choices in public and private sector partnerships promise to provide for sustainable poverty reduction and human development.[2] In this context, World Heritage sites have become demonstration projects where local community engagement has informed sustainable development. This case study presents Hoi An Ancient Town in central Viet Nam.[3]

Hoi An has more than 1,390 architectural remains representing ten architectural forms. These are residential houses, places of worship for family ancestors, village communal houses, pagodas, temples, tombs, bridges, water wells, markets and assembly halls. The foundations of Hoi An go back to the late 16th century. However, the present architectural remains are mainly from the 19th century. There is evidence of the economic and cultural exchanges between Viet Nam and the world, between Hoi An and China, Japan, India and several European countries. The residents of Hoi An, generation after

---

[1] Executive Director, International Institute for the Inclusive Museum, Denmark.

[2] UNDP Viet Nam, 1998, *Expanding Choices for the Rural Poor, Human Development in Vietnam*; UNDP, 2010. *Human Development Report*, New York, United Nations Development Programme.

[3] Hoi An Ancient Town in Quang Nam province, 30 km south of Danang city, was inscribed on the World Heritage List in 1999, under criterion (ii) as an outstanding material manifestation of the fusion of cultures over time in an international maritime commercial centre; and criterion (v) as an exceptionally well-preserved example of a traditional Asian trading port.

Cam Kim island has been famous for boatbuilding. It catered to the transport needs of Hoi An and also to faraway places such as Hue and Ho Chi Minh City. The rapid growth of World Heritage tourism has led to an increased demand for boatbuilding and improved livelihood for the 4,300 islanders.

generation have been living and working in the same houses. In the Ancient Town, a large number of antiquities are kept; traditional crafts are maintained; and folk dishes, habits, customs, and festivities are fostered.

Forging partnerships for sustainable development

The boundaries of the Ancient Town and the surrounding District were established under the Cultural Heritage Law of Viet Nam and the Hoi An People's Committee Statute on Managing, Preserving and Utilizing the Hoi An Ancient Town.[4] To promote responsible conservation and development of Hoi An, the municipal government has divided the town area into two zones. Zone I or Intact Protection Zone covers the heritage in the Ancient Town and the associated areas, which must be protected with authenticity and integrity of the outstanding universal value. Zone II or Ecological Environment and Landscape Protection Zone is the area surrounding Zone I, where structures can be built that contribute to the promotion of the values provided that these

[4] The Master Investment Project For Conserving And Utilizing The World Heritage Site of Hoi An Ancient Town was proposed and ratified by the Prime Minister under Decision No. 240/TTg, dated 14/4/1997.

Hoi An administration provides infrastructure and training for the employment of people confronting the challenges of the legacies of war and disability. Products from surrounding craft villages are marketed to visitors through targeted affirmative action programmes.

109

structures do not affect the architecture, natural scenery and ecological environment of the Ancient Town. Zone II is further divided into Zone II-A and Zone II-B, each with its own set of detailed regulations pertaining to physical interventions of constructions and new developments.

The Hoi An Centre for Monuments Management and Preservation (hereafter HACMMP) drafted a five-year plan in 2002 on the sustainable development of Hoi An to safeguard the outstanding universal values.[5] Four key challenges were identified by the local authorities: conserving the authenticity and integrity of the Ancient Town; meeting the needs of the present residents who actually live in the heritage buildings; promoting and safeguarding the outstanding universal value in tourism development; and improving the income and standard of living of the people without compromising the site's values.

A ten-point Sustainable Developmental Action Plan was adopted by the Hoi An District and Quang Nam Governments:

1. Addressing the organic historical linkages and relationships between Hoi An Ancient Town and the surrounding stakeholder communities and villages as a priority.

[5] Hoi An Five Year Plan, 2002–2007, UNESCO Hanoi and HACMMP, drafted by Amareswar Galla, technical adviser, UNESCO Hanoi.

Hoi An is an outstanding material manifestation of the fusion of cultures over time in an international commercial port. It is an exceptionally well-preserved example of a traditional Asian trading port. Rigorous planning and supporting legal regulations help to safeguard the site's outstanding universal value.

2. Investing more resources from tourism revenue to assist with further restoration of monuments and heritage houses and urban infrastructure development.

3. Implementing affirmative action programmes to ensure the participation of disadvantaged women and young people from the surrounding villages.

4. Improving interpretive materials and cultural tours as part of responsible tourism growth.

5. Developing new attractions in the hinterland to diversify heritage experiences for dispersing rapidly growing visitation and ensuring income redistribution to surrounding communities.

6. Monitoring environmental impacts with the rapid growth of tourism and the increased resident and business activities.

7. Minimizing the rapid degradation of the riverine system through improving infrastructure to deal with commercial waste, sewage from the town and villages and storm water runoff.

8. Controlling floodwater spillage into the town and villages and the consequent damage to heritage buildings and landscapes.

9. Integrated local area planning, both short term and long term, by the local government to continue to enhance the urban infrastructure to cope with development pressures.

Wood craft is the main occupation in Kim Bong village on Cam Kim island. It was historically significant in the construction of wooden houses in the Ancient Town and is now critical for the conservation of the built heritage. Viet Nam National Administration of Tourism led the national pilot project here on community-based tourism and local craft production as part of the export-led Poverty Reduction Programme.

10. Developing an appropriate spatial plan for the whole Hoi An district to continue to resolve the tensions between conservation and sustainable development.

The above principles were embedded into the budgetary process of the governments concerned. The Home Owners Association, Women's Union and Youth Union provided the civil society participation in drafting the Action Plan. A system of tax concessions among commercial street businesses was introduced to deal with disparities in wealth generation from the rapid growth of tourism. A total preparedness and response mechanism was introduced to address environmental degradation and disaster preparedness. Thus the Action Plan became the strategic stakeholder partnership committed to locating outstanding universal value in sustainable development.

Appropriate management

The Action Plan was harmonized with the governance and management of the Ancient Town. In addition to legal documents such as the regulations on managing, conserving and utilizing Hoi An Ancient Town, the municipal

The assembly halls and communal houses in Hoi An were built to serve the Chinese community. They incorporate Chinese religious and architectural elements conforming to the principles of *feng shui*, and integrate architectural and stylistic elements from Vietnamese building traditions.

government has issued regulations on business, advertisements and environmental hygiene. Project investment has been managed in three categories:

- Architectural structures: classifying the historic buildings into different categories according to proprietary forms, historical and cultural values, conservation levels; using the classification as a basis to apply effective conservation and restoration methods.
- Infrastructure development: restoring and upgrading the systems of water supply and drainage, electricity, transportation; preventing environmental pollution and preserving natural landscapes.
- Capacity-building: training for heritage management staff and artisans from craft villages; conducting archaeological surveys for the purpose of restoring and preserving heritage assets.

Developments were regularly monitored, inspections were carried out, and stakeholder community participation was ensured. A major evaluation carried out in 2008 that underlined community engagement and ownership has had positive effects on the conservation of Hoi An.[6] Reliable and rigorous data management was introduced so as to inform and ensure appropriate planning

---

[6] Report on the Results of Implementing the Master Investment Project for Conserving and Utilizing Hoi An World Heritage Site – Quang Nam Province (under Decision No. 240/ Ttg, dated 14/4/1997 by the Prime Minister), 2008.

Table 1 Growth in visitation at Hoi An

Category of visitors	2005	2010	2015	2020
Domestic	350 275	615 496	1 138 915	2 128 232
International	342 859	736 648	1 377 377	2 469 134
Total visitors	693 134	1 352 144	2 516 292	4 597 366

Table 2 Overnight stays or bed nights in Hoi An

Category of visitors	2005	2010	2015	2020
Domestic	57 820	128 800	271 700	545 000
International	289 082	568 800	1 071 300	1 929 000
Total lodgers	346 902	697 600	1 343 000	2 474 000

Table 3 Accommodation unit/room forecasts in Hoi An

Accommodation	2005	2010	2015	2020
Standard rooms	2 703	4 744	10 336	19 832
Non-standard rooms	30	32	35	35
Total	2 733	4 776	10 671	19 867

Table 4 Labour forecasts for Hoi An tourism to 2020 (unit: person)

Category of labour	2010	2015	2020
Direct labour in tourism	5 056	11 459	26 772
Indirect labour in services	11 122	25 210	53 544
Total	16 178	36 669	80 316

Table 5 Income forecasts from Hoi An tourism to 2020 (US$ million)

Category of turnover	2010	2015	2020
Domestic visitors	25.99	85.51	190.75
International visitors	134.41	1071.3	2314.8
Total turnover	160.40	1156.81	2505.55

Table 6 GDP norm and investment capital forecast for Hoi An tourism (US$ million)

GDP	2010	2015	2020
Projected growth in Hoi An town	195.5	596.7	895.1
Projected growth of An tourism and trade	104.26	358.7935	613.143
Rate of growth of tourism and trade against the GDP projections for Hoi An town	53.33	60.13	68.5
Anticipated investment demand for tourism growth	312.8	1076.4	1962.048

and governance. This approach to planning through information management is of particular significance at this World Heritage site, given that the total area of the Hoi An district is about 60 km² and the local population is only about 80,000.

The rapid growth of visitation and the urgency of measures for sustainable development of the site are evident from the baseline data of 2005. The data in Tables 1 to 6 are provided and cross-checked by the HACMMP. The source is Hoi An Trade and Tourism Department.

In this context of rapid growth, the development of Hoi An District, conservation of the Ancient Town and the transformation of management at HACMMP for ensuring outstanding universal value in sustainable development is driving relationship-building with local communities. It has four elements: promoting a shared understanding of heritage conservation and sustainable development through projects; bringing together local communities and villages to take ownership of heritage conservation and outstanding universal value; diversifying visitor experiences through interpretive activities; and integrating conservation in economic development and responsible tourism. What has become critical is building a sense of ownership among local community groups through projects that demonstrate direct community benefits. The following examples illustrate some of the projects.

A former chairman of Hoi An People's Committee once famously pronounced that without the World Heritage Ancient Town, Hoi An would die.

114

Enabling stakeholder community benefits

In the first decade after World Heritage inscription, approximately 200 government-owned heritage buildings were restored at a total cost of more than US$5 million. The municipal government provided 45 per cent of the total funding, while the national and provincial governments contributed 50 per cent. Donor support accounted for 5 per cent. About 1,125 privately owned heritage

Hoi An lamps have become a major craft product for both domestic use and sale in pro-poor craft tourism. Young people and women are especially supported in lamp production through training and associated infrastructure assistance.

115

buildings were repaired by the owners according to the restoration permits that were issued after the owner submitted a plan and budget. Because the cost of restoration of historic buildings is high relative to the income levels of most of the owners of heritage buildings, the municipal government provided partial subsidy for several private conservation projects. Financial assistance was based on the classification of buildings according to heritage values, location and the economic situation of the owner.

Annual investment on the restoration of seriously damaged old houses has helped families with financial hardships to commit themselves to protecting the heritage of the Ancient Town. The municipal government provided three-year loans without interest. In some cases the local government purchased privately owned heritage buildings from families who have economic difficulties and want to sell their houses. After renovation the previous owners could continue to live in the same place at a reasonable rent. This has prevented outside interests from purchasing the properties and has enabled residents to remain in their homes. Entrance fees to the Ancient Town contribute to funding these interventions.

In the last century due to wars and famine, and more recently due to the introduction of plastic and other mass-produced goods, centuries-old traditional craft villages were fast disappearing. These villages were an organic part of the Ancient Town, located in its immediate hinterland. The revitalization of these villages, the use of their skills in the conservation and restoration projects of the Ancient Town, and initiatives to alleviate poverty around the World Heritage site were prioritized. The villages and the industrial area outside the Ancient Town catering to the building industry were rezoned with buffer areas to minimize negative impacts on the outstanding universal value.

Thanh Ha, a riverine village, was established towards the end of the 15th century as a ceramic village. It provided, especially from the 17th to the mid 20th centuries, products catering for Hoi An construction and utilitarian needs. Innovation and creativity even led to the export of some wares to other provinces and overseas. In the late 20th century, plastic and metal tools flooded the market, contributing to the rapid decline of the village. It is now rehabilitated with the craftspeople working again on the conservation and restoration of the Ancient Town. Their skills have also diversified so that they are able to meet the unprecedented demand for bricks, tiles, decorative and other materials in the hotel industry.

Kim Bong woodcraft village caters for the complex waterways of the region requiring boat building customized to local needs. It is revitalized, providing for restoration work and tourism development. Groups of carpenters are also able to come together in the production of handicrafts. New technologies are used along with traditional skills to increase productivity and make the workplace safer. Fishing and tourist boats fitted with modern equipment and meeting the demands of visitor safety are also being made in Kim Bong.

Tra Que horticultural village is about 2.5 km north of the Ancient Town. For the past four centuries, generations of villagers have lived and worked in this specialized village catering for the herbal and vegetable markets in the Ancient Town. This may be one of the few heritage villages with cultural values derived from the days of the Cham civilization. The main developmental objectives were to safeguard the traditional horticultural knowledge and to develop interpretive experiences for local school children and visitors. Rehabilitating the environment of the village, promoting scientific research on biodiversity, and establishing a youth camp have contributed to the historical and environmental experiences for visitors.

Vong Nhi, an ancient fishing village, is 2 km east of the Ancient Town in the Cam Thanh commune, which is an integral part of the Cham Island Biosphere Reserve. Revitalization of the village has enabled the continued service to the well-known fish market in the Ancient Town. New measures have been introduced to minimize environmental impacts on the hinterland of both the village and the Ancient Town. Research and development has led to the reclamation of the historical significance of the fishing village and the salvage of heritage sites and landscapes in the face of commercialization through aquaculture.

The distribution of heritage tourism benefits throughout Hoi An Ancient Town has been part of the spatial planning strategy for the sustainable development of the site. Bach Dang Street and its river frontage, for example, specialize in the local culinary heritage, and the intangible heritage of Hoi An is showcased there on full moon nights.

117

Interventions have resulted in employment and regular income for fishing families; preservation of the elements of a traditional fishing village; new visitor experiences; interpretation of fishing heritage, including sites, material culture and folk life; and raising local consciousness about environmental conservation.

In addition to the historic villages, the hinterland of the Ancient Town has a significant array of inscriptions and archaeological remains covering a period of almost two millennia, from the prehistoric Sa Huynh to the famous Cham civilization. These have been systematically mapped using digital technologies and interpreted. Visitors are able to appreciate the historical context of World Heritage through an introduction to these remains at the four conserved houses converted into museums in the Ancient Town. A new cultural centre and social history museum was opened in 2009 to provide the much-needed information and orientation facilities that are critical for visitor management. Collections, exhibitions and public programmes are planned to promote an appreciation of the outstanding universal value of Hoi An Ancient Town.

In the Ancient Town, infrastructure improvement includes installing underground systems of electricity, telephone, cable TV, water supply, fire control, roads and sidewalks. Repairs to the Hoi An market helped the local people and suppliers from the surrounding villages to continue to use the space that is at

The revival of traditional festivities has enhanced both the visitor experience and the local communities' sense of place. These include harvest festivals, dragon boat racing, and public prayers for prosperous fishing.

least two centuries old. The gradual building of dykes and dredging the riverine system is reducing the impact of floods. The construction of an underground system of collecting solid waste and wastewater has made most houses habitable.

The revival of traditional festivities has enhanced both the visitor experience and the local communities' sense of place. These include harvest festivals, dragon boat racing, and public prayers for prosperous fishing. The full moon festival is a new introduction. Electricity is turned off and the Ancient Town is lit with Hoi An Lanterns. Performances and concerts derived from the intangible heritage of Hoi An are organized in both the restored houses and on the waterways. In addition to the commercial benefits to the local communities, these festivals have become an important mechanism for dispersing visitors and dealing with congestion in the Ancient Town while promoting the outstanding universal value.

Learning and sharing experiences

The LEAP project provided valuable opportunities for learning from and sharing with comparative World Heritage sites such as Lijiang Ancient Town in China.[7] Valuable lessons have been learnt from the project-driven approach to conservation and interpretation in Hoi An. In order to effectively manage, conserve and promote a heritage site, it is essential to have comprehensive and long-term strategies based on conservation principles and community interests and benefits through strategic stakeholder cooperation, including governments, scientists and researchers; and the heritage house owners, business operators and other local people.

The revitalization of Nguyen Thai Hoc Street is an exemplary demonstration project. Visitor services in 2000 were concentrated in Tran Phu and Le Loi Streets. This led to a large gap in income between the home owners in these streets and those in other streets. In order to encourage an even spread of visitation and income, a revitalization programme was initiated focusing on the development of tourism services along Nguyen Thai Hoc Street. Funds have been mobilized to restore twenty-six ancient houses in this street with twenty used for residential and commercial purposes, two as exhibition centres and museums, two as offices, one as a communal house and another as a traditional craft workshop. Homeowners in the street were granted licences for businesses such as tailoring shops, art galleries and souvenir

---

[7] Integrated Community Development and Cultural Heritage Site Preservation Through Local Effort in Asia and the Pacific (LEAP). http://www.unescobkk.org/culture/world-heritage-and-immovable-heritage/leap/leap-projects/

Thanh Ha ceramic village kilns dating back to the 15th century have provided construction materials for World Heritage projects on rehabilitation, conservation and restoration work in Hoi An Ancient Town. Villagers' skills have also been diversified to meet the unprecedented demand for bricks, tiles, decorative materials and crafts in the tourism industry.

shops, which were previously restricted to Tran Phu and Le Loi Streets. Improved equality in income distribution between homeowners has also provided the property owners with incentives and resources to maintain their heritage buildings.

Synergies in conservation

Ten years after the inscription of Hoi An Ancient Town on the World Heritage List, in 2009 the adjacent Cham Island Marine Protected Area was recognized as a Man and the Biosphere Reserve by UNESCO. Cooperation and coordination with all the stakeholders to maximize the benefits of conservation and minimize the negative impacts is a shared commitment between the two UNESCO listings. Local authorities, conservation agencies, civil society, the business sector, and local communities participate in the management of the outstanding universal value and biosphere. This has been tested and proven to be productive in the previous decade with the Hoi An Ancient Town.

In building on the demonstration projects of the previous decade, there are many challenges that are envisaged in the new Master Plan for 2020 that brings together both the Ancient Town and Cham Island. The focus on the Ancient Town will be extended with a greater emphasis to the surrounding hinterland. New sources of investment will be solicited. Annual flood mitigation demands major solutions. The majority of the houses built with wood have to be protected in the face of fire damage, a constant threat. Termite infestation is also an ongoing problem. Rapid growth of tourism calls for new strategies for visitor management and decongestion of the Ancient Town. While many heritage homeowners are now able to finance the restoration of their houses, the pace of rebuilding has to be monitored and controlled. The capacity of management staff and government officials has to be kept abreast of rapid growth. The carrying capacity of the Ancient Town is being reviewed, with all motorized traffic already restricted in 2012. While a number of projects have been developed to disperse visitors across the district, sourcing investment in activities outside the Ancient Town has been a challenge.

119

In 2008, Hoi An was chosen as one of the 110 historic destinations in the world by *National Geographic* magazine. Hoi An, along with its hinterland the ancient Faifo, was one of the most significant localities of the Cham civilization, informing the new cultural route including the nearby My Son and Hue World Heritage sites. The focus on the livelihood of the primary stakeholder communities using World Heritage as a powerful tool for articulating their unique sense of place and identity has become the lifeline for the sustainable development and growth of Hoi An and the region. The core strategy is to minimize the negative impacts of rapid growth while endeavouring, through a range of demonstration projects and tax measures, to distribute the growing wealth of the area to eradicate poverty and destitution.

10 Responsible local communities in historic inner city areas: Historic Centre (Old Town) of Tallinn, Estonia

RIIN ALATALU[1]

Historical meeting point

Tallinn Old Town Conservation Area was established in 1966. The Historic Centre (Old Town) of Tallinn was listed as a World Heritage site in 1997 under criteria (ii) and (iv) along with many other historic towns. The conservation governance framework was already established by the time the World Heritage Convention was drawn up in 1972. Although considerable work on protecting the history and heritage of the capital was done at specialist level as early as the 1960s, the involvement of the local community in protecting and evaluating heritage has developed in parallel with the Convention and informed by growing universal consciousness and responsibility for heritage conservation.

Tallinn World Heritage site is the heart of the busy capital of Estonia. Sustainable development of this living city depends on diverse factors, principles and even aftermaths of historical events. Tallinn is a significant meeting and interchange point of different cultures – religions, lifestyle, building traditions and nationalities. Estonia since the Crusades of the 13th century has been a multinational and multicultural society. Safeguarding the diverse values, authenticity and integrity in the urban area made it imperative for the heritage authorities and local communities to take on considerable responsibility, demonstrate flexibility, broad scope and diverse actions.

The stakeholder community is locally understood as a group of interacting people living in a common location and sharing some common values.

---

[1] Head of the Division of Milieu Areas, Tallinn Cultural Heritage Department, Estonia.

Old Town Society's campaign to remind drivers of special traffic restrictions in the Old Town.

The local World Heritage site community is based on respect for the internationally and domestically recognized values of a unique and authentic historic city.

Community consists of numerous different subgroups and parallel groups formed on different principles. The local World Heritage site community is based on respect for the internationally and domestically recognized values of a unique and authentic historic city. In Tallinn there are several different communities – local societies, non-governmental organizations, cultural and educational centres and also the city authorities, who all have their role in the sustainable development of the Old Town.

Management of a living World Heritage site is a complex matter, cooperation and understanding between different responsible parts is essential. This case study focuses on the rich history of Tallinn Old Town and its protection and presents some good examples of community involvement in safeguarding the site's outstanding universal value and shaping a living environment that is enjoyable for all.

A living town

Tallinn received City Rights in 1248 and soon after it became a member of the Hanseatic League – one of the most powerful trading organizations in Northern Europe during medieval times. The uniqueness of Tallinn is not only the remarkable extent of preserved medieval urban fabric of narrow winding streets, many of which retain their medieval names, fine public and burgher

Archaeological investigations have shown that the Harbour of Tallinn on the Viking route to Constantinople, has existed since the 10th–11th centuries.

123

buildings, including the town wall, churches, monasteries, merchants' and craftsmen's guilds and the domestic architecture of the merchants' houses. Tallinn is unique for having a balanced combination of architectural layers from different centuries bearing witness to the times of glory and decline. The importance of Tallinn is that it has preserved its authentic historical layers of significance while developing as a modern and functioning urban centre.

Many of its districts and buildings are adapted to contemporary variations of historical functions. For example the upper town, Toompea, houses the Government and Parliament of Estonia and many embassies. It has always been the administrative centre of the country. The only preserved Gothic Town Hall in Northern Europe has functioned as the ceremonial building of city government since 1404. Tallinn is also proud of the fact that a pharmacy opened in 1422 is one of the oldest drugstores still operating in Europe. The Old Town is home to 4,000 people and still serves as a trading centre. Maintaining historical functions is essential not only for the sustainability, but also to support the key elements of outstanding universal value and to promote the long traditions of Tallinn.

Sustaining heritage values

One of the crucial issues for Tallinn has been to keep the Old Town functioning as a heart of the capital. As in many other World Heritage sites, Tallinn

Middle Age Day. Tallinn has held traditional Middle Age Days since 2000 alongside several other traditional events promoting the history and heritage of the city.

124

confronts the tensions between heritage conservation needs and the aspirations of the local communities that co-exist. In a rapidly globalizing world that is market driven, it has become important to define priorities and values and to share them with the society to secure a common ground to safeguard the outstanding universal value.

It is considered that World Heritage conservation is first of all a responsibility of the local community. In a city such as Tallinn one of the most important tasks is to keep local people coming to the Old Town on a daily basis. This has been possible through civic spaces such as theatres, cinemas, concert halls, museums and restaurants within the Old Town, augmented by the location of the government, parliament, four ministries, municipal boards, several university departments and five secondary schools in the Old Town. Working in the city or visiting offices, accompanying their children and grandchildren to school, has given people a personal touch with the Old Town and raised awareness of the historic heritage.

Working in the city or visiting offices, accompanying their children and grandchildren to school, has given people a personal touch with the Old Town and raised awareness of the historic heritage.

Historic areas are often under pressure from tourism, and this is the case in Tallinn. There has been considerable discussion about moving public offices out of the Old Town, the most expensive premises in Estonia, and turning historic buildings into profitable hotels and tourism-oriented enterprises. In maintaining the historic surroundings, local authorities, entrepreneurs and inhabitants

Tallinn is an outstanding and exceptionally complete and well-preserved example of a medieval northern European trading city.

125

have endeavoured to ensure quality experiences and best impressions for visitors and the tourism industry. But the key question for sustainable maintenance and development of authentic heritage is still how locals value the Old Town and what has to be done to promote responsibility for common heritage.

Tallinn has a tradition of Old Town Days since 1982, Heritage Month 18 April to 18 May since 1985, European Heritage Days since 1991, Middle Age Days since 2000, and several other traditional events promoting the history and heritage of the city. Dressing up for carnivals a couple of times a year is important for promotional purposes, but it is not enough to raise the awareness of the community. Continuing education and promotion has become essential. The main focus in Tallinn is on guided tours and lectures to promote history and explain the heritage. In 2011 a new campaign, Back to School!, was launched during which heritage specialists give voluntary lessons in schools.

Outstanding universal value management

In the early 1950s the first municipal department for the maintenance of Tallinn Old Town was formed, and Tallinn Cultural Heritage Department has existed for more than half a century. Although the name and structure of the department have changed several times, it has administrated historic premises based on several legislative and local government documents protecting the values of Tallinn Old Town. The most important documents are the Statutes of the Heritage Conservation Area of Tallinn Old Town, first approved in 1966, and the Heritage Conservation Act. The Statutes regulate the maintaining of the historic plot structure, building volume and density, historic structures and details of the World Heritage property in the Conservation Area and its buffer zone.

Appropriate legislation is important as it guarantees the common understanding of restoration principles for all parties involved. Approval for any changes in historic premises needs to be based on knowledge and experience and to serve heritage values. Since Tallinn was inscribed on the World Heritage List, the administration contract between Tallinn City Government and Estonian National Heritage Board was concluded. According to the contract, decisions concerning planning and building within the World Heritage property are made by consensus of the National Heritage Board and Tallinn City. Specific questions or decisions of utmost importance which need even wider consensus are taken to the Heritage Conservation Advisory Panel, which consists of our best independent experts.

Following the advice of the World Heritage Committee, the Tallinn Old Town Management Committee was established in 2010. Maintenance of a living city depends on numerous stakeholders and different challenges. The aim of the Management Committee is to strengthen cooperation and coordination among responsible organizations, NGOs, local community and other stakeholders. It is also responsible for approving, enhancing and monitoring implementation of the comprehensive Management Plan for the property.

A responsible local community

At the time the World Heritage Convention was drawn up in the 1970s, international responsibility for natural and cultural heritage was growing. The reasons for this were different in various states, as very often heritage is raised into public focus at times of social, environmental or political crisis. For occupied Estonia heritage was a question of national pride. Since the end of the 1960s the broad-scale movements of nature protection societies and later country study and heritage protection societies started to influence the common understanding of heritage values.

Panoramic view of the Old Town. Despite the decline of the Hanseatic League from the 15th century, fine public and domestic buildings continued to be built in Tallinn according to prevailing architectural taste.

Estonia is well-known for its high respect for heritage. In the late 1980s, heritage societies were the ideological leaders in the movement for independence and memory and historical values were the key symbols in rebuilding the national state. The changed society of the 1990s led to altered relations in heritage management. After regaining independence and integration into the market economy, heritage policies faced a crisis – people cared more for modernizing society than taking care of historic environments, and were more involved with their personal careers and new challenges, so community values dropped into the background. The relaxation of control by the state, but also by the community over inexperienced new private owners, caused much harm to the listed monuments.

Since the early 2000s the society has stablilized and interest in heritage has again become a common concern. With the support and social control of the community, the state and municipal institutions regained control over restoration practices, not only in the Old Town and listed monuments but also in other historic buildings and regions of Estonia. Dozens of local societies have been formed to participate in the preservation of values of the living environment. People gather to get to know their neighbours, but also to take care of the surroundings, organize meetings with the representatives of City Government, keep an eye on or even intervene in planning processes, inform the authorities about neglected houses and so on. The Tallinn Old Town Society, for example, unites the local residents in conserving the site's outstanding universal value.

An engaged local community

Estonians are traditionally very enthusiastic about education and educational activities. One of the best-known traditional movements in Tallinn is the youth club Kodulinn (Hometown). Since 1975 schoolchildren have gathered every second Sunday to help the conservators and to clean and maintain Tallinn Old Town. The effect of these youngsters' voluntary work was multiplied by a popular TV serial promoting the movement. They were also involved with Old

Estonia Song Festival. In maintaining the historic folklore, local authorities, entrepreneurs and inhabitants have endeavoured to ensure quality experiences for visitors and the tourism industry.

Town Days and published their own newspaper. For decades more than 10,000 youngsters have received their first heritage education in Kodulinn. A number of present-day politicians and well-known people have been involved with the movement and demonstrated their commitment. In addition to the voluntary physical work, the club continues its educational activities.

> Heritage is about people. The cultural value of any World Heritage site will only be maintained if citizens are ready, able and willing to inherit it and if they consider this inheritance to be of value.

One of the best-known examples of heritage-based education is Collegium Educationis Revaliae (CER) – a school in Tallinn that was created on the initiative of parents. The aim of the school is also to be an innovative cultural and social centre that functions as a community space. Being community based means that students, teachers, graduates, parents, grandparents and friends participate actively in educational and cultural life, and in this way build up the centre. The number of students has been growing constantly during the twenty-five years of CER's existence and today there are around 1,000 students. CER combines general and targeted education in heritage, arts and music. Much attention is also paid to students with special needs. The school community participates in the building of the Old Town of Tallinn as an active cultural and living environment and contributes to cultivating heritage awareness. Keeping the Old Town meaningful and open to local people has become a crucial issue today to avoid the threat of turning a living city into a commercial tourist area.

The church of St Olaf is in typical basilical form, with lofty vaulting and a precise geometry of form in what is recognized to be the distinctive Tallinn School.

CER follows the idea that there can be no heritage without intergenerational transmission. Heritage is about people. The cultural value of any World Heritage site will only be maintained if citizens are ready, able and willing to inherit it and if they consider this inheritance to be of value. For this reason the CER team has, from the very beginning, aimed to build up an educative environment in Tallinn Old Town that has an impact on the students of the school. Renovation of the High School building, Vene 22, which is an outstanding historic monument, served as the best opportunity for this purpose. The architecture of an educational institution is at the same time education in architecture, history, culture and art history. The well-restored historic schoolhouse teaches the history of Tallinn from the Middle Ages to the present day, because all major changes in history can be seen and studied in this environment. CER has premises in several historic buildings in Tallinn Old Town that are closely connected, enliven the whole area and give it a special spirit. Among them the High School building is worth a special mention. From the end of the 13th century until the Reformation it belonged to Cistercian monasteries. After the Reformation the plots were for a long time in private hands. The school was started in one part of the building in 1800, and the whole place has served different schools until the present day. In 1996 Vene 22 was given to CER by the city administration. The building was in a very poor state as it had been neglected for over fifty years and in the 1990s only occasional emergency repair works were carried out. During this period the school board started to work at turning the house into a study environment and exposing its rich historical background.

In 2004 a new stage of renovation began with investigations. In 2005 the special requirements of restoration were prepared by the Tallinn Cultural Heritage Department and the architectural renovation project was also finalized. In 2006 the Tallinn City Government decided to support the renovation and the actual period of restoration lasted from January 2007 until January 2008. The objective of the restoration was to explore, open and expose the layers of centuries of history in the neglected schoolhouse carrying the stigma of the carelessness of the fifty-year Soviet era, and to show the different functions of the complex throughout history. The main principle of the work was respect for the aesthetic and historic values of the building. The goal of conserving as much as possible and restoring as little as possible secured the preservation of the authentic constructional fabric, allowing the building to expose all its historic layers in this day and age. After the renovation the schoolhouse has become a place of study with modern facilities set in historic surroundings.

The school community has also helped in this work. For example, the cellars were emptied and cleaned with students' help. The community has also organized several fund-raising activities. For about fifteen years traditional charity fairs have been held twice a year where everyone can make their contribution. This can take various forms, for instance one could bring a handmade item for sale, bake a cake, give a concert or donate a piece of art for a charity auction. All the proceeds are used to further improve the school environment for the children.

In 2010 CER was one of the founding members of the Our Heritage foundation together with an NGO, the Estonian Heritage Society and the Estonian National Commission for UNESCO. The aim of the foundation is to encourage citizens to safeguard our common heritage and to be actively involved in decisions on how it should be used. The founding members consider it important that buildings of heritage value should be given a function that respects heritage, promotes outstanding universal value and maintains the town as a living environment. A big charity concert was organized by CER students, teachers, graduates, parents and friends and its proceeds were donated to the fund.

CER is an active member of the UNESCO Associated Schools Network and has implemented several projects at local and international levels to promote World Heritage values. In short, it has managed to create a truly sustainable environment that forms a perspective for a meaningful future. Thanks to the school, the parents, grandparents, graduates and friends form a well-functioning network. The CER buildings constitute an environment that helps children and their families to make personal contact with the historic townscape and heritage. The school workshops are open to all visitors to the city. Only when the Old Town has a meaning and value to local people, will it be a source of information and inspiration to visitors.

Only when the Old Town has a meaning and value to local people, will it be a source of information and inspiration to visitors.

130

From the local to the universal

While the World Heritage Convention celebrates its 40th anniversary, Tallinn celebrates fifteen years on the World Heritage List. World Heritage status has given Estonia a broad scope and the opportunity to compare its values and also protective measures with the rest of the world. Being part of a big picture has helped in finding solutions in troubled times and enhancing delight over success stories.

Estonia is proud of its community awareness and involvement in heritage conservation. It is also aware that beneficial cooperation of authorities and communities needs constant engagement. Without mutual support, authorities and communities become rivals instead of partners. A major breakthrough towards the sustainable development of Estonian heritage started when the heritage authorities made a serious about-turn in their attitude towards publicity, themselves contributing to the voluntary work promoting heritage, explaining and teaching.

Estonia has a long-lasting tradition of voluntary community work. The anniversary year of the Convention is also the year of World Cleanup 2012 – a global initiative started by Estonia to clean our planet. The main aim of the campaign, with already more than eighty countries participating, is to educate people and promote a sustainable and ecological lifestyle.

Home is where the world begins. Big issues start at the local level. Respect for local heritage is essential for developing respect for World Heritage and making the world respect yours. This is the shared ethic that has become integral to the commitment of Estonia to the outstanding universal value of the Old Town of Tallinn.

Respect for local heritage is essential for developing respect for World Heritage and making the world respect yours.

131

11

An exceptional picture of a Spanish colonial city: Historic Centre of Santa Cruz de Mompox, Colombia

JUAN LUIS ISAZA LONDOÑO[1]

500 years of heritage values

This is a very significant colonial settlement with more than 500 years of layered history and heritage values virtually intact, together with an intangible heritage legacy that holds potential and a firm relationship with the economic activities of its residents.

The Historic Centre of Santa Cruz de Mompox is the result of a set of complex associations and diversity of heritage. It is located on the natural landscape featuring a highly swampy territory, a branch of the Magdalena River, the country's most important waterway. This is a very significant colonial settlement with more than 500 years of layered history and heritage values virtually intact, together with an intangible heritage legacy that holds potential and a firm relationship with the economic activities of its residents.

The rise and stagnation of this site as a commercial port was due mainly to the change of the main stream of the Magdalena River which led to a decline in its commercial role, indirectly resulting in the conservation of its cultural heritage built over centuries. Today the economic stagnation is reflected in the poverty indicators of its population, not only at the site itself but throughout the region, which has also been affected by floods, especially during the last two rainy seasons.

In the fight against poverty, the protection of natural and cultural heritage has become the focus around which priorities are set and a chain of actions are taken that enable not only the preservation of the site's outstanding universal value but also its social appropriation and sustainable development by the community, including economic and social development. In the long term, the educational, environmental, cultural, aesthetic and social aspects of the conservation of the World Heritage site will be more important than its economic impact. However, the people of Santa Cruz de Mompox face to a large extent the needs, difficulties

[1] Director of Heritage, Ministry of Culture of Colombia.

Mompox cemetery.

and challenges of the economic conditions of developing countries, which make it necessary to analyse heritage conservation from the standpoint of its opportunity to generate a positive economic impact.

In 2011, the UNESCO Resource Manuals included a series of directives to ensure that any use of the World Heritage sites was sustainable and ensure the safeguarding of the outstanding universal value and to reaffirm, as a non-negotiable principle, the idea that the management systems of the sites must integrate sustainable development principles. These directives must be applied in the case of the municipality of Santa Cruz de Mompox to preserve the outstanding universal value of the Historic Centre, where the protection and conservation of the cultural heritage is coordinated with actions that respond to the environmental, social and economic needs and with active and continuing participation by the community.

Environmental, social and economic characteristics

The branch of the Magdalena River, the axis along which the historic centre was shaped, is one of its main natural attributes. The site is also located in a territory where 70 per cent of the municipal area corresponds to water from marshes formed by the confluence of the country's main rivers in a zone known as Depresión Momposina.

The river is the axis of the natural heritage of Mompox, of the recreational activities, of the fauna, the vegetation and the climate which is characterized by high levels of humidity and temperatures around 40 degrees Celsius. The most common trees in the region – *Ceiba bruja*, jagua, oak and palma de vino – have a large dense canopy and are the habitat of many bird species such as the guacamayo, as well as a good source of shade. These trees are generally found in the patios and backyards of the buildings in the Historic Centre, among which the rubber tree at the Hostal Doña Manuela stands out. This area is also home to wildlife such as howler monkeys, squirrels, iguanas and lizards, as well as birds including cattle egrets, swamp cat, kingfisher, vultures, macaws, frigate birds and pelicans. This ecosystem enriches the Historic Centre's cultural value and holds potential both for scientific research as well as tourism specializing in this type of heritage.

The area of the Historic Centre inscribed on the World Heritage List is 30 per cent of the urban zone, so preserves its main functional role for the residents, who according to the 2005 census numbered 41,565 in the entire municipality, of which 54.4 per cent live in the urban area. This tangible and intangible cultural heritage is closely connected to a natural environment dominated by the river and other waterways. However, this wealth contrasts with social and economic needs. In Colombia, the most common indicator to measure poverty is that of Unsatisfied Basic Needs (NBI: Necesidades Básicas Insatisfechas).[2] The 2005 census indicates that 51.6 per cent of the population have NBI (32.9 per cent of the urban population and 74.2 per cent of the rural population). Thus the basic needs are yet to be met of almost 7,500 people in the urban area and 14,064 people in the rural zones who have a strong connection with the Historic Centre of Mompox. In addition, of the 15,724 people deemed to be economically active, 17.2 per cent lack education, while 38.2 per cent have received primary-school education and only 9.4 per cent technical or higher education.

The region's economy is based on extensive cattle farming, growing of natural pastures, small-scale agriculture of citrus and other fruits, industrial and artisanal fishing; commercial use of forest species; provision of services and low-scale production of arts and crafts and, to a lesser extent,

[2] The NBI methodology seeks to determine, with the help of some simple indicators, the basic needs of the population. Groups that do not reach a minimum set threshold are classified as poor. The indicators selected are: unsuitable housing, households with critical overcrowding, houses with unsuitable services, households with high economic dependence and households with school-age children who do not attend school, National Administrative Department of Statistics (DANE: Departamento Administrativo Nacional de Estadística): http://www.dane.gov.co/index.php?option=com, 2008.

Bell tower of San Augustin church.

tourism. The main barrier to the development of agricultural activities is the difficulty of commercializing the products due to accessibility problems, lack of support prices, high intermediate costs, the absence of agricultural credits for small producers, regulations, poor technology and inadequate research, and the scarcity of training for farmers. Fishing activities are also affected by difficulties in the navigability of the river and the environmental pollution that has caused disease and the disappearance of several fish species.

The economy of the city is particularly dependent on the production and commercialization of goods and services for local consumption. According to the 2005 census, 54.2 per cent of the economic activities in Mompox are of a commercial nature and 15.6 per cent are industrial. The commercial activities produce low income levels. A survey conducted in 2008[3] during the formulation of the Special Management and Protection Plan of the Historic Centre of Santa Cruz de Mompox indicates that of the people that perform exclusive economic activities in the city centre, 79 per cent pay rent, 60.6 per cent have waste collection services, only 20 per cent recycle, 51 per cent have a telephone connection and only 9.6 per cent an internet connection. The business premises are very small, as is their economic impact: 86.17 per cent of them have less than three employees, monthly sales for 27.7 per cent are under US$250 (less than the current minimum wage) and for 29.8 per cent sales are between this amount and US$500. The approximate monthly income for 87.5 per cent of people is below US$250. Among the self-employed, 73.7 per cent have had a business for a period ranging from one to five years. There are no records about how many *momposinos* are engaged in informal trade but in the main streets of Mompox it is very common to find this activity, especially the sale of prepared food.

The industry is carried out at artisanal level and knowledge is passed from generation to generation, enriching the intangible heritage of the population and maintaining a unique art of carpentry, cabinetmaking, woodcrafting,

135

---

[3] The survey was conducted in 216 businesses of the Historic Centre, of which 94 are exclusively dedicated to this economic activity. Ministry of Culture, 2009, Special Management and Protection Plan of the Historic Centre of Santa Cruz de Mompox.

Schoolchildren enter the church of San Francisco which served as a fort in the early years of the settlement.

gold- and silversmithing (notably filigree work), ceramics and pottery, weaving, embroidery, forging and blacksmithing, calabash engraving and the production of sweets and fruit ciders.

The unique production of gold and silver pieces of high aesthetic value emerged from the mix of Arab techniques with the pre-Columbian tradition of goldsmiths from the Zenu area, who used the melted wax technique to manufacture the filigree threads to create nose rings, chest pins, stick heads and a variety of zoomorphic shapes. In Mompox, this long tradition is promoted today by two associations in goldsmiths' workshops which are often in the house of the craftsman or workshop owner, with more than four goldsmiths under his responsibility. The main difficulty for these workers remains the lack of commercialization of their product, although it is already recognized at national level.

All these traditional crafts are part of the intangible heritage of the site, which in recent years has served as an educational and promotional centre for traditional crafts aimed at disadvantaged young people.

All these traditional crafts are part of the intangible heritage of the site, which in recent years has served as an educational and promotional centre for traditional crafts aimed at disadvantaged young people. The School Workshop Santa Cruz de Mompox, supported by the Ministry of Culture and the Spanish Agency for International Development, is an institution like those in most countries of Latin America that seeks the integration of students in the labour market through conservation and cultural heritage promotion projects. In Mompox it is also one of the few options for the repair and restoration of the built heritage to be carried out by those with knowledge and experience of traditional construction techniques.

Tourism only takes place during determined periods of the year and is very intermittent due to access difficulties, the shortage of public utilities and the poverty of the population, among other circumstances.

In summary, the entire population presents a list of needs almost as long as the list of attributes that inform the outstanding universal value of the site. The Municipal Development Plan – Future Vision 2004–2007, which was one of the sources of the diagnosis that enabled the formulation of the Special Management and Protection Plan of the Historic Centre indicates, among others, the following issues:

- Environmental deterioration accelerated with the degradation and loss of natural resources. Quantitative and qualitative deficit in health and education services, as well as in housing for low-income families, in sport, recreational and cultural services.
- Increase of unemployment indexes, accelerated increase of the informal economy.
- Deterioration of the urban and rural road network and lack of public spaces.
- Public security issues and low community participation.
- Need of institutional strengthening for the municipal administration.

137

Special Management and Protection Plan

Colombia's Culture Act sets forth the need to formulate a Special Management and Protection Plan (PEMP) as the main instrument for planning and management of cultural sites such as the Historic Centre of Santa Cruz de Mompox. This plan must specify the area affected, the zone of influence, the intervention level permitted, the management conditions and the dissemination plan to ensure community support for heritage conservation.

The Special Management and Protection Plan of the Historic Centre of Mompox was approved by the Ministry of Culture by Resolution 2378 dated 17 November 2009. The purpose of the measures specified in the PEMP is to guarantee the conservation and sustainability of the Historic Centre and seek to produce an economic, social and environmental impact and, along with that, contribute to the sustainable development of the community. This plan, in addition to defining the possible interventions to bring the buildings up to modern standards of comfort, also includes a series of themes and strategies that work towards resolving the general problems of the population, as indicated in Table 1.

Table 1 Themes and strategies contributing to sustainable development

Themes	Strategies
Essential aspects: attention to fundamental needs	Assurance of good accessibility Regional integration and strengthening Improvement of quality of life
History and heritage of the cultural traditions and artistic expressions	Protection and rescue of the historical memory Promotion of cultural expressions, traditions and tangible and intangible heritage
Economy, entrepreneurship and productivity	Entrepreneurship development Strengthening of association capacity: productive chains and associative businesses Marketing and promotion Promotion of research, development and innovation
Cultural, competitive and sustainable tourism	Mompox as a competitive tourist destination Mompox as a sustainable destination
Built heritage: old sector as integral group	Structured actions Actions from the systems: networks Protection and revaluing of the architectural heritage Planning and zoning
Institutions, management and dissemination	Institutional strategy Management and funding strategy Dissemination plan, communications and participation

Outcomes for local communities

Several of the proposed strategies were addressed by the ICOMOS report at the time of inscription of the site on the World Heritage List in 1995, such as recovery of the river's importance, restoring the paving and development of a detailed tourism plan.

From the urban intervention viewpoint, five major projects inform the recovery of public spaces:

- Plaza Fundacional de la Concepción.
- Mompox de Cara al Río: Eje de La Albarrada or Carrera 1ª (includes social projects and the relocation of residents occupying risk areas on the banks of the river).
- Gathering and meeting axis: Calle 18.
- Calle Real del Medio Symbolic axis.
- New city axis: Carrera 5.

Local children returning from a festival.

These are actions not only at the World Heritage site but also in the buffer zone given the strong functional, social, economic and spatial relation between the two parts of the city. All the projects include physical intervention in squares, parks, streets and sidewalks and architectural intervention projects for the conservation of buildings, as well as new material to improve the infrastructure and provide the services needed by the residents, among which are the main churches, a support centre for citizens' participation and a new hospital.

With regard to the ecosystem, recovery and management programmes of the environmental protection zones have been formulated, the most important being the Magdalena River branch, the Bosque Santander and the Botanical Garden, the vegetation of parks, squares, roads and backyards of buildings. At regional level, there is a need to integrate the ecologic structure of the region and risk management due to flooding.

The PEMP also proposes architectural interventions by way of general plans that involve several buildings at the same time, such as the restoration of abandoned buildings, façades, roofs and backyards. The need to promote technical training of the population for such interventions is also indicated.

The tourism development plan being put together is based on the carrying capacity of the site, with an emphasis on the prevention and elimination of negative impacts.

The main responsibility for implementing the PEMP rests with the Mayor's Office in coordination with the Ministry of Culture, which in addition has to advance the institutional strengthening process, dissemination of information

Sunset along Magdalena River.

and community participation. The Mayor's Office is also required to present semi-annual reports on the implementation of the PEMP.

Colombian legislation has been very ambitious and precise in the relation that must exist between preservation of the cultural heritage and the social and economic development of the population, the task still being open to advances in instruments and policies to enable synergies between natural and built heritage. In fact, one of the management conditions in National Decree 763 of 2009, which regulates to a large extent the tangible cultural heritage of the nation and stipulates the mandatory formulation of a PEMP for urban areas that have been declared properties of natural interest, refers to the financial aspects, whose guidelines and determiners must seek the preservation and sustainability of the cultural good. Decree 763 states that the financial aspects entail: 'Economic and financial measures for the recovery and sustainability of the good of the property, that comprise the identification and formulation of projects to incorporate the same to the economic and social dynamics and to determine the sources of resources for their conservation and maintenance. It incorporates the tax aspects regulated by this decree' (Article 21, Section iii).

In executing this mandate, the PEMP of the Historic Centre of Santa Cruz de Mompox specifies the following economic measures: quality of life, participation and cultural expressions; comprehensive management programme for solid waste, improvement and expansion of the sewage system, waste water treatment plant, improvement of aqueduct service; school of dance, music and traditional chants; road, river and air accessibility; improvement of health services;

Colombian legislation has been very ambitious and precise in the relation that must exist between preservation of the cultural heritage and the social and economic development of the population.

During Holy Week families honour the memory of the dead. Funerary rites are among those preserved by the Special Management and Protection Plan approved in 2009.

strengthening citizen participation. Proposals concerning tourism include the creation of a specialized touristic land and river transportation service and a tourist offer portfolio connected to cultural and natural heritage, the restoration of the river ports of Magangué, La Bodega and Mompox, plus staff training. To develop other economic activities, the need to act on training is being addressed by business loans and the promotion of associations. The financial sources identified for the execution of the PEMP correspond to the municipality's resources and the contributions from regional and national government.

The main purpose of the PEMP is that it guarantees conservation and sustainability of the World Heritage site to produce economic, social and environmental impacts and thus contribute to the sustainable development of the community. The mission of the Colombian state by supporting local government in the implementation and monitoring of the plan is to verify the expected impact, both in the preservation of the integrity and outstanding universal value of the cultural heritage, as well as its contribution to sustainable development. Taking as reference Donovan Rypkema's five quantifiable measures of the economic impact of heritage conservation given in a presentation to the 2005 European Cultural Heritage Forum,[4] (1) employment and family income; (2) revitalization of city centres; (3) cultural heritage tourism; (4) property

[4] D. D. Rypkema,. 2005, *Cultural Heritage and Sustainable Economic and Social Development*, European Cultural Heritage Forum, Brussels, Belgium. DRypkema@PlaceEconomics.com

values; and (5) incubation of small businesses, we can indicate in the first place that, although the PEMP was approved in 2009, it is a short period in which to assess its impacts.

Current priorities are the institutional capacity of the District Administration and the impact of the rainy season floods on the population and the country. A large part of the residents suffered the effects of the floods, which also damaged the built heritage and the entire regional economy. The Ministry of Culture has provided technical and financial support to restore the Albarrada,[5] which is still in progress and whose impact will be measured soon. The unprecedented rainfall, that has affected the entire country, caused damage to the retaining wall, requiring emergency consolidation work which received support from the Netherlands Prince Claus Fund's Cultural Emergency Response. The rainfall affected the planned timelines for recovery of the public space and the wall.

The sustainability imperative

The World Heritage inscription is a catalyst for sustainability and for the population to be able not only to appreciate and value their history and heritage, but also to enjoy them in improved quality of life.

In conclusion, in the Historic Centre of Santa Cruz de Mompox, the preservation of the cultural heritage not only can but must contribute to environmental, social and economic sustainability of the population because only in this way can its integrity and outstanding universal value be preserved, the very values for which it was inscribed on the World Heritage List. The positive benefits for the town of Mompox include the recovery of the public space of the Plaza de la Concepción, with the sustainable preservation of the cultural heritage, such as the public market, an organized community of artisans, the increase in sales in shops along the Boroa and in the Plaza, and the improvement of the environmental conditions of the site for residents and users, as well as interventions in response to the vulnerability of the river banks.

The World Heritage inscription is a catalyst for sustainability and for the population to be able not only to appreciate and value their history and heritage, but also to enjoy them in improved quality of life. The priority has been strengthening the capacity of the population and of the local administration, with emphasis on the relationship between cultural, natural, economic and social heritage in an environment in which strong relations with the entire region coexist. The most important lesson is that when governments are transient, the most important way to ensure sustainable outcomes is through the strengthening of the local community and of its capacity to contribute and demand integral measures for improvement in quality of life, where cultural heritage is the main driving factor.

---

[5] The Albarrada is the river's retaining wall or dyke, as well as the name of the town's public space facing the river.

3

Integrated Planning and Indigenous Engagement

Planning has often been perceived from a formal administrative viewpoint. However, knowledge communities working at local level in the conservation of heritage values and their physical manifestations at sites and landscapes have used their cumulative knowledge over generations to make choices and strategic approaches in managing various cultural and environmental resources. Local and indigenous peoples are now universally recognized for their systematic approach to decision-making both in preservation of their heritage and the transmission of heritage values to the next generation, as demonstrated by the five case studies in this chapter.

Indigenous peoples numbering over 35 million are the custodians of significant cultural and biological heritage. The 2007 United Nations Declaration on the Rights of Indigenous Peoples clearly calls on respecting the first voice and decision-making of indigenous people in addressing their well-being and caring for their unique but highly fragile knowledge systems. UNESCO is committed to prioritizing the cultural and linguistic diversity of indigenous peoples and respecting their valuable knowledge systems.

Sustainable development of the Sacred Mijikenda Kaya Forests (Kenya) is establishing the important role of local communities in the Kayas (clearings), the control of access to sites and the safeguarding of their intangible heritage. Although they do face many challenges, the Kayas have demonstrated the power of traditional knowledge, practices and communal commitment in the protection of World Heritage.

In Canada's Jasper National Park, the Jasper Aboriginal Forum has enabled significant reconciliation and reconnection initiatives with the Aboriginal groups. Each of the participating communities has been involved in cultural renewal, healing and seeking greater self-determination and economic opportunities. The outcomes have become significant for the World Heritage site.

Sustainable development of Chief Roi Mata's Domain (Vanuatu) is ensuring the continuity of the outstanding universal value of the site and

Local villagers overlooking plains from the Cliff of Bandiagara. The sanctuary lies at the southern limit of the Sahara in an arid Sahelian region with averages of 580 mm of rainfall per year.

in doing so working through primary stakeholders to find pathways for future development of the local communities so that the benefit sharing does not diminish the inheritance of the indigenous peoples. The selection criteria for the inscription of this site succinctly underline the intergenerational responsibility that is the essence of sustainable development.

The Rice Terraces of the Philippine Cordilleras are a product of tradition and the role of the cooperative systems of the primary stakeholder, the Ifugao communities, is at the heart of their creation and maintenance. The Ifugao, as the civic agency instrumental in ensuring regulated economic progress, are the lynchpin guaranteeing sustainability of the territories within and adjoining the rice terrace clusters, the culture and traditions, and the natural environment in its political, social and economic context. They are the drivers of World Heritage conservation.

At the Cliff of Bandiagara (Land of the Dogons) site (Mali), the importance of sustainable development of a living site such as Dogon Country requires community participation and promotion of the local economy. The local development projects initiated by various partners are all opportunities for properly aligning economic and social interests with cultural interests, and this includes the conservation of the site itself.

12 Homelands of the Mijikenda people: Sacred Mijikenda Kaya Forests, Kenya

GEORGE OKELLO ABUNGU AND ANTHONY GITHITHO[1]

Abodes of ancestors

The Mijikenda Kaya Forests (*Ma-Kaya* in the Mijikenda language) are located on the Kenya coast. They belong to the closely related linguistic groups known collectively by the same name, Mijikenda. The main topography of the Kenya coast includes a flat plain edged by sandy and coral cliffs as well as mangrove swamps, from which a range of low sandstone hills rises to a maximum height of about 250 m parallel to the coastline. From these hills there is a drop, sometimes precipitous, to the Nyika plateau followed by a gradual descent to the semi-arid and flat Taru desert.

The forested sites are on hilltops and sometimes in valleys in this landscape, mostly in the sub-humid coastal range and on the plain itself. They are typically found in the midst of densely populated rural farmlands dominated by coconut and cashew-nut tree stands and clusters of thatched dwellings, in the homelands of the Mijikenda people.

The contrast between the surrounding farm monoculture and the luxuriant indigenous forest groves is strong and the *Ma-kaya* stand out conspicuously, mysterious and alluring. They appear undisturbed, but all true Kaya forests bear the clear imprint of humanity. They have visible clearings at their centre and a system of deeply incised and well-worn paths leading to and from these spaces. In some of the clearings, there are stands of coconut trees indicating past settlement. From the air, this consistent pattern of paths and clearings in the *Ma-kaya* is particularly striking. The effect is that of land sculpture with the collective

---

[1] George Abungu is the former Director-General of the National Museums of Kenya;
Anthony Githitho is a Senior Research Scientist with the National Museums of Kenya.

Kaya Rabai. The Mijikenda Kaya Forests consist of ten separate forest sites spread over some 200 km along the coast.

148

The effect is that of land sculpture with the collective 'sculptors' being the local rural Mijikenda people who through time have been inspired by their changing environment.

'sculptors' being the local rural Mijikenda people who through time have been inspired by their changing environment.

Kaya forest groves range in area from 30 ha to 300 ha and are the remains, preserved by cultural norms, of formerly much more extensive forests. To date, over fifty Kayas have been identified in the contiguous districts of Kwale, Msambweni, Kinango, Kaloleni, Mombasa, Kilifi and Malindi of the coastal region of Kenya. Most Kaya forests tend to be located at strategic sites on hill-tops but a few are found in river valleys and others on flat land. The type of vegetation of the Kayas varies depending on the type of forest or woodlands that originally dominated the area.

By definition, Kayas differ from other types of forest, and even other sacred places of the Mijikenda, in having a history or tradition of settlement of the site, which is also closely related to the myth of origin and migration of the different Mijikenda communities. The marks of organized human activity are associated with their ancestors who lived there in the past – 'Kaya' literally means 'home' in most Mijikenda dialects.

Mijikenda people

The Mijikenda people are the dominant ethnic community in the coastal region of Kenya between the border with the United Republic of Tanzania

Rabai Kaya Elders in procession. The Kayas were traditionally managed by the Elders, who defined norms, practices and taboos for the care of these sites.

in the south and the northern limit of Malindi district near the Tana River. This is a strip of over 300 km with a varying width of between 50 km and 60 km. They are in fact nine distinct groups (Mijikenda means 'nine tribes') but speak closely related Bantu dialects which share about 70 per cent of their vocabulary, suggesting that their separation and formation as different groups may have begun less than 1,000 years ago (Nyamweru, 1998, pp. 8–9). The groups are: A-Giriama, A-Digo, A-Duruma, A-Chonyi, A-Ribe, A-Rabai, A-Kambe, A-Kauma, A-Jibana (with the prefix 'A' denoting a people or tribe).

Mijikenda oral history states that the ancestors of the Mijikenda, who were then one people, lived in a place called Singwaya, believed to be north of Tana River and south of Juba River in Somalia. However due to conflicts with other communities there they migrated south in waves into the present Kenya coastal region from the early 16th century onwards (Spear, 1978).

They established themselves in fortified villages known as Kayas. As they continued to be harassed by other groups, especially nomadic pastoralists, the defensive function of the Kaya village was crucial to their survival. This was achieved by siting a village within thick forest so that it could only be approached on narrow forest paths, surrounding it with a sturdy stockade, or burying within the Kaya the sacred objects (*fingo*) essential to the material and spiritual well-being of the community (Nyamweru, 1998). The Kaya forests with their clearings and sacred sites are what remain of the formerly extensive forests and hidden villages, preserved now as ritual and spiritual sites, the surrounding land having been given over to agriculture during the 19th and 20th centuries.

The Kaya forests today therefore comprise distinct cultural landscapes closely linked to the traditions and history of the coastal Mijikenda communities. They display consistent cultural elements which may be regarded as diagnostic and include:

- association with the tribes' migration from Singwaya;
- presence of the protective charm *fingo* or talisman from Singwaya;

149

Lwanda ritual hut in central clearing. The Kaya forests, with their clearings, are what remain of the former forests and hidden villages, preserved now as ritual and spiritual sites.

- a defined Kaya, or central clearing;
- Moroni: the site of the Moro or historical meeting hut;
- Mwara (plural *Nyara*) paths;
- Mvirya (gate) sites;
- Makaburini or burial sites;
- Chiza, prayer sites or altars.

Kaya forests and natural diversity

The coastal forests of Kenya are part of a regional system of forests extending as far south as Mozambique, known as the Eastern Africa Coastal Forests Ecoregion, which includes the Zanzibar-Inhambane Regional Mosaic (White, 1983). The forest region exhibits a very high level of biodiversity in terms of sheer diversity, endemism and rarity in a significant number of biological groups. Part of the reason for this is geological. Most of the present geomorphological features of coastal East Africa have developed over the last 200 million years. The entire range of geological substrate found in Africa

Visitors walking with Kaya Elders in Kaya Rabai. All have to subscribe to rituals before being allowed into the Kaya, including the practice of taking off their shoes.

is present in the coastal forest belt from pre-Cambrian (around 2,500 million years old) rocks to recent alluvial deposits. Forests are found on plains, plateau, marine and lacustrine deposits. All these factors contribute to the great diversity of vegetation types and the effect is further heightened by the variety of climatic regimes and soil types.

As part of this system and remnants of what is believed to have been much more extensive forest along the Kenya coast, it is not surprising that Kayas display high biodiversity values in terms of diversity, endemism and rarity. The latest estimates show that Kayas constitute about 5 per cent of the remaining coastal closed forest cover of Kenya, estimated to be about 67,000 ha, yet when an assessment of plant biodiversity values was made, seven out of the twenty sites with the highest conservation status were Kaya forests.

This status has been further heightened by human activity over the millennia. Human dispersal and migration coupled with population growth led to forest clearing for settlement and agriculture. With the gradual abandonment of the central Kaya settlements over the last two centuries it is believed that the Mijikenda cleared much of the primary forest in the surrounding areas where they live today (Spear, 1978). The result is that the forest areas containing the Kayas have become more and more isolated as their connection with each other has been lost. Forest islands have formed with their resultant rich biodiversity.

Kayas and World Heritage

The Kaya forests are not only unique in their nature and history but also in their place as hot spots of biodiversity, including traditional medicinal plants. They are an enclave of culture and nature in otherwise highly populated areas with great commercial value, being adjacent to some of the most precious coastal lands. The obvious values of these sites and their

151

Shrine in a typical forest clearing at Rabai Forests. The Kayas are now the repositories of spiritual beliefs of the Mijikenda and are seen as the sacred abode of their ancestors.

152

strong interrelation of cultural and natural components prompted Kenya to present them for consideration as World Heritage. In 2008, ten Kayas were listed as a World Heritage site having outstanding universal value under the following criteria:

Criterion (iii) The Kayas provide focal points for Mijikenda religious beliefs and practices, are regarded as the ancestral homes of the different Mijikenda peoples, and are held to be sacred places. As such they have metonymic significance to Mijikenda and are a fundamental source of Mijikenda's sense of 'being-in-the-world' and of place within the cultural landscape of contemporary Kenya. They are seen as a defining characteristic of Mijikenda identity.

Criterion (v) Since their abandonment as preferred places of settlement, Kayas have been transferred from the domestic aspect of the Mijikenda landscape to its spiritual sphere. As part of this process, certain restrictions were placed on access and the utilization of natural forest resources. As a direct consequence of this, the biodiversity of the Kayas and forests surrounding them has been sustained. The Kayas are under threat both externally and from within Mijikenda society through the decline of traditional knowledge and respect for practices.

Criterion (vi) The Kayas are now the repositories of spiritual beliefs of the Mijikenda and are seen as the sacred abode of their ancestors. As a collection

Typical contemporary houses of villagers living near Kaya Kambe.

of sites spread over a large area, they are associated with beliefs of local and national significance, and possibly regional significance as the sites extend beyond the boundaries of Kenya.

Kayas in sustainable development

Sustainable development refers to economic growth in which *resource* use aims to meet human needs while preserving the *environment* for present and future generations. While economic welfare is commonly implied, the term is not restricted to it. In examining the Kayas' contribution to sustainable development it is important to note their utility for local people and the wider society. This is directly linked to their outstanding universal value and primarily consists of, although not limited to:

- traditional, cultural values for local Mijikenda, group identity and sense of spiritual well-being and 'being-in-the-world';
- natural heritage conservation, areas of natural habitat and species conservation;
- local environmental services, soil conservation, springs, etc.

All the above are clear developmental needs, although not explicitly economic. While there is material consumption of the forests locally in terms of poles, thatch and firewood, this is not encouraged traditionally or by national law and there is no suggestion that this is a primary use and benefit of the Kaya resource for local people. The same goes for direct economic benefits although there are clear linkages and the Kayas provide a supportive environment. There is remarkable consensus on this point locally, nationally and even globally. It follows that in pursuing sustainable development there is a need to build on or develop the primary values of the Kayas, while conserving the natural resource itself, in other words limiting physical extraction.

This is reflected in the type of economic or livelihood projects which have been developed in relation to the Kayas to improve economic conditions, which have tended to be in two main areas:

- Cultural / ecotourism. Visitation of Kaya sites by non-Mijikenda under controlled conditions, as at Kaya Rabai and Kaya Kinondo.
- Nature-based enterprise projects linked to the site but not extracting from it. The main activity in this respect has been bee-keeping at sites such as Kaya Giriama, Kaya Rabai and Kaya Kauma.

A potential economic opportunity which would fall into the second category is the emerging carbon trade where sites are able to attract payment due to their carbon-fixing ability and environmental services. A pilot project is under way in Dzombo, a forested hill containing a Kaya on the south coast of Kenya.

Safeguarding the Kayas

Traditionally, the Kayas were managed by the Elders. They defined norms, practices and taboos for the care of these sacred sites, thus the laws governing their protection and the associated rituals all remained intact. Today however the management frameworks for Kayas are characterized by collaboration, the main partnership in site protection being between the local communities (Mijikenda) and the National Museums of Kenya. The sites themselves have been designated National Monuments under national law.

An important role of local communities (Mijikenda) in Kayas is control of access to the sites and custody of the intangible cultural heritage.

All the Kayas in the World Heritage site have management plans developed by the National Museums of Kenya and the Kaya Elders and communities and their joint management can be said to represent some of the best practices. An important role of local communities (Mijikenda) in Kayas is control of access to the sites and custody of the intangible cultural heritage.

Local communities also manage enterprise projects, either directly or through oversight groups. Through the Museums and other bodies such as non-governmental organizations, the communities are linked with enterprise

Information Centre at Kaya Mudzimuvya, one of the Rabai Kaya. Cultural ecotourism has been one of the main factors contributing to the improvement of economic conditions.

155

partners. This makes use of the respective strengths of different groups and the communities have received donor funding for projects from WWF and the French and American embassies in Kenya, among others. At Kaya Rabai, for example, a visitor centre has been constructed and local people engaged in relevant activities such as bee-keeping for sustainability, demonstrating that with support, communities can be engaged in profitable activities through the use of their heritage.

Community benefits

The main objective of managing the Kayas for sustainable development therefore is to conserve the sites for continuation of their primary benefits for local people and wider society. These by definition are often non-monetary, but crucial for social well-being and healthy development, further assuring the identity of the Mijikenda community as well as the protection of their culture, environment and knowledge systems.

The livelihood projects mentioned above have realized modest returns. The relatively small area of the Kayas would in any case be insufficient to make a significant impact on the local economy through productive extractive land-use. However their significance and importance in cultural, spiritual and heritage terms is enormous, although no systematic computation has yet been made of this. The Kayas are used regularly for traditional purposes by local Mijikenda; they have a high profile on the local political scene in the coastal region, and the status they have gained in the public eye as an indigenous African World Heritage site both locally and nationally is very notable. These factors more than make up for the relatively small direct economic benefit as a contribution to sustainable development.

Valuing traditional knowledge

The Kayas are a good example of a successful and implementable traditional management system that similar sites can borrow from. Furthermore they offer a case study of successful co-management of community and government bodies with support from NGOs and other donors. As an institution, they demonstrate the capacity of the communities to plan, maintain and utilize resources in a sustainable way in a World Heritage site.

The other area of interest is in resource mobilization and use, where the Kayas have demonstrated that the community can mobilize labour for various activities, as well as the Elders' contribution to knowledge accumulation and enhancement by working with researchers in the identification and use of plants and other resources, mostly through oral forms of intellectual engagement. This has further contributed to the appreciation and retention of the traditional knowledge system within the society and in the site management.

It has been demonstrated through the Kayas that it is not only possible to have co-management by two different institutions, one governmental and one traditional, but that both a written Management Plan as well as undocumented traditional management systems can be applied without conflict in the successful and effective management of a World Heritage site.

Continuing challenges

Although they do face many challenges, the Kayas have demonstrated the power of traditional knowledge, practices and communal commitment to the protection of heritage.

Despite the successes described there is still a major challenge in providing economic opportunities for local people, especially youth. The Kayas are first and foremost sacred places and cannot be freely opened up to the public, including tourists. Thus visitation is restricted and so are the opportunities for maximization of economic benefits.

There have been developmental pressures, as some of these Kayas are located on prime lands that attract the attention of property developers. Many Kayas have been taken and turned into new development properties against the wishes of the communities.

There is population pressure and the need for land for agriculture, wood resources and even a breakdown in traditions that has seen some people steal and sell cultural items, such as decorated grave posts from the sites, as antiquities to collectors. In some instances, some Kayas have become victims of mineral exploration and exploitation as well as sand harvesting for the thriving building industry along the coast, to accommodate tourists as well as private house buyers.

Due to the location of many sites on relatively high ground compared with the surrounding areas, some of the Kayas have become the site of telecommunications towers. Fortunately many of the Kayas, including all the World Heritage ones, are declared national monuments protected by Kenyan laws and as such have not become victims of these vagaries. Although they do face many challenges, the Kayas have demonstrated the power of traditional knowledge, practices and communal commitment to the protection of heritage. They are indeed a good example of community contribution and responsibility to a heritage site and of commitment to ensuring its outstanding universal value through a proactive co-management system – a good lesson to all who care for the heritage of humanity.

157

Two sacred grave posts (*vigango*) brought back from the United States after many years are put back in their rightful place.

13

Reconnection and reconciliation in Canadian Rocky Mountain Parks: Jasper National Park, Canada

CYNTHIA BALL, SHERRILL MEROPOULIS, AMBER STEWART AND SHAWN CARDIFF[1]

The cultural importance of a remarkable landscape

Jasper National Park is the largest of Canada's southern mountain national parks at 11,228 km^2 and part of the UNESCO Canadian Rocky Mountain Parks World Heritage site. It supports Rocky Mountain mammals, including caribou, grizzly bears, wolves, bighorn sheep and mountain goats. It protects the headwaters of several major rivers including the Athabasca, a Canadian Heritage River. It exemplifies the outstanding physical and biological features of the Rocky Mountain Biogeographical Province. The Burgess Shale Cambrian fossil site in Yoho National Park and the Precambrian fossil sites in Mount Robson Provincial Park contribute important information about the Earth's evolutionary history. Classic illustrations of glacial geological processes are evident throughout the site, including ice fields, remnant valley glaciers and exceptional examples of associated erosion and deposition features.

The Canadian Rocky Mountains are renowned for their scenic splendour, which includes striking mountain vistas, glacial lakes, alpine meadows, vibrant local mountain culture and diverse sightseeing and recreational opportunities. More than 95 per cent of the site remains as undisturbed natural wilderness and there is a high measure of ecological integrity throughout, with strong legal protection, which contrasts significantly with the modified landscapes in other parts of the Canadian Rocky Mountains. Jasper National Park hosts 2 million visitors annually.

[1] Parks Canada Agency, Jasper, Alberta, Canada.

Tipi encampment at the Moberly Homestead, Jasper National Park. The establishment of Jasper National Park in 1907 triggered social changes for the Aboriginal peoples.

159

Jasper's Athabasca Valley bears evidence of 9,000 years of human use. Archaeology and oral histories confirm that the lands that are now Jasper National Park were crossroads and 'commons' shared by people of the forests, plains and mountains. Due to Jasper's location, groups representing a rich blend of cultural traditions had come from both sides of the continental divide, travelling to obtain specific resources, to gather for trade and to conduct ceremonies. No one Aboriginal group claimed exclusive use of the area. Perhaps due to limitations of climate, soils and food availability, there was no known permanent occupation of the park area by Aboriginal peoples[2] until the arrival of the fur trade in the early 19th century and the establishment of the Métis homesteads in the 1890s. Until recently, little attention has been paid to the cultural importance of this remarkable landscape to the Aboriginal communities that have historic and contemporary ties with it. This case study discusses how one of the component protected areas – Jasper National Park – has embarked on a significant process of reconciliation to re-establish broken cultural ties and reintegrate Aboriginal uses and perspectives into its management.

[2] The term 'Aboriginal peoples' is a name given collectively to the original peoples of Canada – First Nations, Inuit and Métis – and their descendants (Constitution Act, 1982).

In 2010, representatives of Jasper National Park participated in a healing ceremony with the Alexis Nakota Sioux First Nation. Indigenous peoples have long practised renewal ceremonies.

Parks and Aboriginal peoples

The establishment of Jasper National Park in 1907 led directly to the severing of links with Aboriginal peoples who had historically used the area. Lands taken up for the new national park were part of the traditional territories of a number of First Nations groups who may or may not have signed treaties (1876–1898), and six Métis families who had recently established homesteads. While the treaty-making process included provisions for lands to be taken up in the future for other uses – in this case, conservation and tourism, it remains in doubt whether affected Aboriginal peoples were ever consulted about the establishment of Jasper National Park.

The Métis families were financially compensated, but they and other Aboriginal peoples were not allowed to continue practising their traditional resource harvesting and cultural practices in the park. Aboriginal peoples had little choice other than leave, taking with them their stories, cultural traditions and intimate knowledge of the area.

What occurred here and across western Canada reflected policies designed to open up land for settlement and development. There were profound social consequences for Canada's Aboriginal peoples as a result of the pursuit of this objective and the loss of traditions. Many took their cultural practices underground. We now know that in Jasper National Park, many Elders quietly continued their ceremonies, teachings, and gathering of medicines, discreetly, under a veil of secrecy and in fear of prosecution.

The ceremonial pipe was raised in 2006 by groups attending the inaugural Jasper Aboriginal Forum, as a powerful symbol of unity, solidarity and a lasting promise to work together in good faith. The pipe remains a symbol of the spirit of reconnection, the goal that brought all parties together. The ceremony and subsequent meeting marked the end of a century of virtual silence between Aboriginal communities with historical connections to Jasper National Park and the Parks Canada Agency responsible for its management.[3]

[3] Parks Canada's mandate is: 'To protect and present nationally significant examples of Canada's natural and cultural heritage and foster public understanding, appreciation and enjoyment in ways that ensure their ecological and commemorative integrity for present and future generations.'

Harvesting Chief Root, a traditional medicine. Indigenous Elders who maintain the teachings of their ancestors share the gift of their knowledge of medicinal plants.

Each of the par-
ticipating communi-
ties was involved in
processes of cultural
renewal, healing,
and seeking greater
self-determination
and economic
opportunities.

161

Each of the participating communities was involved in processes of cultural renewal, healing, and seeking greater self-determination and economic opportunities.

They were experiencing chronic land use pressures and environmental impacts in their traditional territories from industrial developments of many kinds. Regional Aboriginal communities have long held the mountains and headwaters to be a sacred place for refuge and healing, and the national park offers the assurance of clean air, pristine waters and a healthy landscape, with protection in perpetuity. These trends led to the strong interest shared among some twenty groups in reconnecting with Jasper National Park.

Parks Canada Agency is now at the forefront of the Government of Canada's ongoing efforts to reconcile with Aboriginal peoples. The cultural shift within the organization was enabled through policy and legislative changes to the Canada National Parks Act (2000). These changes opened up the use of park lands and natural objects by Aboriginal peoples for traditional spiritual and ceremonial purposes. In many cases, new park establishment occurs with the support of Aboriginal peoples and under agreements that entrench traditional uses and terms for co-management. Parks Canada welcomes Aboriginal peoples as partners in the protection of natural and cultural resources, and recognizes that in order to successfully implement our mandate, we need the counsel of those who used these places historically.

The Jasper Aboriginal Forum grew from these roots as a platform for relationship-building and to reverse the effects of the loss of knowledge, culture and connection. It has succeeded in creating trust through dialogue and as a first step in establishing a productive, collective working relationship between Jasper National Park and more than twenty Aboriginal groups with historic connections to the park. Our challenge in Jasper is to ensure that the cultural needs and aspirations of Aboriginal peoples are considered in a long-established national park with tightly-managed land-use policies and a legal framework never designed or intended to accommodate traditional Aboriginal uses. Yet, the reasons for working together are compelling, and the

Local and regional Aboriginal leaders participate in the raising of the Two Brothers Totem Pole.

Jasper Aboriginal Forum proceeds under the terms of an interest-based process.

Jasper Aboriginal Forum

A journey of partnership and reconciliation began in September 2006 with a letter inviting 'all Aboriginal groups *that we are aware of* who have a historic tie to Jasper National Park to attend' the first Jasper Aboriginal Forum. Earlier, successful collaboration on several economic and cultural initiatives occurred with the Métis Nation of Alberta and representatives of the descendants, which awakened Parks Canada to the diverse Aboriginal history connected to Jasper National Park and the benefits of partnership.

The objectives of the meeting were to learn about the historic connection that different Aboriginal groups have with Jasper, and about Canada's national parks and major initiatives affecting the park, and explore how Aboriginal groups could be involved in future planning.

What began as a conversation to share historical information and identify interests, has evolved into an effective means of achieving specific actions and results.

The Jasper Aboriginal Forum is a biannual business meeting between the Superintendent of Jasper National Park, with the support of park staff, and a political representative and a community-recognized Elder from each of the Aboriginal groups. Gatherings typically focus on the six key areas of common interest identified through group discussions. These conversations are augmented by special presentations and field trips which cover a wide range of topics, from elk management to controlled burning and caribou conservation. They provide information and a basis for knowledge exchange. By extension, the Jasper Aboriginal Forum is also an informal culture camp and a central gathering place for member groups. Over time, trust has developed as participants have learned about each other leading to sharing, teaching and reconnection.

Two Parks Canada Aboriginal Liaison staff coordinate communications and site visits between Aboriginal partners and Jasper National Park, providing the capacity to advance Aboriginal-related initiatives. Community visits are an important extension of the Forum. Parks Canada representatives travel to

162

This sweat lodge was used during the regional indigenous celebration associated with the raising of the Two Brothers Totem Pole. Only the frame remains.

163

Aboriginal communities to discuss topics such as management planning and conservation strategies. These visits require time, commitment and resources and have solidified relationships with community members.

The members of the Aboriginal Forum and Parks Canada have identified six primary areas of common interest:

- providing opportunities to influence park management and decision-making;
- gathering traditional knowledge about historic use of the park;
- facilitating access to and reconnection with the park;
- providing cultural programming to share Aboriginal stories and culture with park visitors;
- participating in resource conservation activities;
- facilitating economic and employment opportunities.

The Forum as a productive partnership

The deliberations of the Jasper Aboriginal Forum and Parks Canada have led to the following results for the first time:

Aboriginal voices are influencing park planning and decision-making

The latest Jasper National Park of Canada Management Plan (2010) contains a key strategy devoted to advancing shared interests. The comments of Elders, community representatives and members were welcomed, carefully consid-

ered and synthesized into the plan's vision statement and contents, which sets
the stage for increased implementation of shared priorities.

Park visitors are enjoying new and inspiring cultural experiences

Aboriginal interpretation is now part of the core offer for park visitors;
interpretive media specific to both Métis and First Nations are either in
place or under development; Aboriginal interpreters provide programming
for understanding indigenous cultures; and residents and visitors immerse
themselves in the rich traditions of First Nations and Métis peoples as they
celebrate National Aboriginal Day in Jasper.

In summer 2011, Parks Canada hosted the largest Aboriginal gathering in
Jasper's history as fifteen regional Aboriginal communities returned to celebrate their traditional ties to Jasper. They danced, drummed, sang and spoke of
their connection to this place for the public. In private, they prayed and passed
the pipe in sweat lodges, honouring their ancestors, and the past, and celebrating new beginnings. Park visitors watched as carvers from Haida Gwaii completed the Two Brothers totem pole. Two hundred people pulled on the ropes
to hoist the pole as more than 5,000 watched. A public feast and Round Dance
completed the day.

Parks Canada has gained insight into Aboriginal ties to the national park

Several Aboriginal communities have shared what they know about past use
of the park and places that hold special cultural significance. An example is
the traditional land-use study carried out as a planning project for the world-
renowned Icefields Parkway.

Reconciliation ceremonies are marking new beginnings

Parks Canada is engaged in a process of reconciliation with the communities
represented at the Forum. For some, this process has involved formal ceremonies, sweat lodges, songs, offerings and prayers of forgiveness, while for others
a renewal of trust is developing through the Jasper Aboriginal Forum, allowing
both the Park and the communities to reconnect.

The 2011 totem-pole raising and celebrations may not have happened without the reconciliation process. In 2008, the minister responsible for Parks
Canada made the decision to commission a new Haida totem pole to replace
the pole that had stood as a landmark in Jasper for ninety-four years, originally
acquired when the railroad was built through the mountains to the sea. The

Mist rising off the Athabasca River, a Canadian Heritage River.

decision was the subject of much discussion and debate as the Haida pole is not culturally authentic to Jasper and not representative of the area. However, with one exception, the member communities of the Jasper Aboriginal Forum took the view that the project could be a positive opportunity to profile Aboriginal cultures and turned this potentially divisive initiative into a powerful event of celebration, reconnection and reconciliation. The groups collaborated with Parks Canada and the Haida Nation in raising the new pole and hosting a series of ceremonies, celebrations, feasting, and a culture camp, memorable for the many Aboriginal and non-Aboriginal peoples who participated.

Former policy barriers are being overcome to allow for new levels of park access

165

To date, five Aboriginal Forum groups have signed bilateral Memorandums of Agreement (MoA) with Parks Canada, providing each group with greater access to the park. This is an interest-based document and is not legally binding. Groups that have signed the MoA can obtain a Jasper National Park Aboriginal entry pass, entitling community members to free park entry for ceremonial, spiritual, religious, cultural or traditional purposes. The pass has created goodwill and recognition that the park is a place of welcome.

Parks Canada has worked with Forum communities to develop a mechanism for harvesting of plants and natural objects for spiritual and cultural purposes.

The pass does not authorize hunting, fishing or trapping within the national park or the harvesting of plants and natural objects for commercial gain. Parks Canada has worked with Forum communities to develop a mechanism for harvesting of plants and natural objects for spiritual and cultural purposes. Parks Canada has also facilitated youth-Elder exchanges where young and old reconnect to the park through traditional teachings.

Parks Canada and Aboriginal communities are working together on the development of a traditional cultural use area

The desire for a dedicated Aboriginal cultural use area within Jasper National Park was identified as a priority at the Jasper Aboriginal Forum. In response, Parks Canada and the Forum membership have agreed on a site that will be a stepping-off place for physical, spiritual and cultural reconnection in Jasper.

Maligne canyon with sun rays. Jasper National Park is one of the components of the Canadian Rocky Mountain Parks World Heritage Site that represents the physical features of the Rocky Mountain Biogeographical Province.

Forum members provide input regarding resource conservation activities in the park

Traditional knowledge is being integrated into a draft conservation recovery strategy for woodland caribou, a species at risk. Cultural informed controlled burning in the park provides an example of how shared interests are served by the Jasper Aboriginal Forum and how communities and park staff are directly engaged in this learning and sharing process. Traditional Aboriginal knowledge strengthens our holistic understanding of ecological integrity issues.

166

Economic and employment opportunities

An excellent example of a successful partnering opportunity is the Métis Nation of Alberta and Parks Canada's Fire Smart Forest Wise employment strategy. This US$900,000 employment and training programme has benefitted Métis youth, supported Parks Canada fire management initiatives and mitigated the destructive potential of wildfire to the town of Jasper.

A tradition of sharing

There is much to be learned in working together with shared goals. The following are some of the core principles for strategic partnerships in working with communities.

Ceremony Understanding that ceremony is at the core of virtually every interaction with First Nations.

Personal integrity and trust A commitment to openness, exchange and support and a willingness to learn about key traditional values, protocols and understandings are crucial elements that have advanced Parks Canada's relationships with Aboriginal peoples.

Capacity and resources Aboriginal communities face many pressures and may have the desire, but lack the capacity and resources, to participate in park programmes. Adequate resources to ensure effective and empowering participation are important.

Be realistic, patient and committed Ensure goals and timelines are reasonable, be committed, and allow enough time for relationships to become established and grow.

Write it down Having a written and signed agreement helps to resolve potential differences. It also helps to maintain the relationship and its customary practices when leadership changes. Documentation of all projects from start to finish is critical so there are photographs, recordings and reference notes for the benefit of future generations.

Ask for Aboriginal perspectives and input early and often Involve people at the beginning of projects and show the work-in-progress to community members. Consulting with all the identified key Aboriginal partners from the outset of a project will garner insight, expertise, consensus and greater participation, leading to a more successful outcome.

Get legal advice When in doubt seek legal advice to avoid jeopardizing the position of either the Crown or the Aboriginal community.

Take it outside Culture camps reintegrate Aboriginal peoples, young and old, to traditionally used lands. Providing access and holding gatherings on the land within a park removes both real and perceived barriers associated with formal meetings and their structures.

Welcome and involve Aboriginal children and youth In most Aboriginal communities, culture and stories are transmitted to children and youth during community events. Promote Aboriginal participation of children and youth in activities, some are even related to more formal processes.

Identify shared values and goals and set expectations Find mutual goals and values that will lead to action plans of mutual interest. Establish clear objectives, and articulate everyone's expectations to lessen conflicting agendas and priorities. Clarify the scope, nature and requirements for consultation.

Today, the Forum's membership has grown to over twenty Aboriginal groups with stated historic connections to the park. The large number of groups involved is a challenge, particularly when attendance fluctuates and representatives change from meeting to meeting. Forum membership remains open – we still have much to learn about the many different Aboriginal connections to the park.

Dialogue and relationship-building

Jasper National Park, part of the Canadian Rocky Mountain Parks World Heritage site, is making significant progress in re-establishing Aboriginal culture and uses in ways that are meaningful to the communities concerned, and supportive of the purposes, protection and celebration of the site. Parks Canada and more than twenty Aboriginal communities began this journey of reconnection and reconciliation in 2006 with the creation of the Jasper Aboriginal Forum, which has provided a platform for relationship-building and created trust through dialogue and action.

Aboriginal voices are influencing park planning and decision-making. Park visitors are enjoying new and inspiring cultural experiences. Youth and Elders are reconnecting through traditional teachings and Parks Canada is visiting Aboriginal communities. We are removing barriers and rethinking how we engage with Aboriginal peoples.

We are learning from each other. We are rethinking the role that Aboriginal peoples played in the park historically, and renewing our perspectives on park ecosystems and ideas of wilderness to reflect a place that people have influenced, and been influenced by, for millennia. We are finding ways to be flexible in accommodating the needs and aspirations of Aboriginal peoples.

The healing process requires time, and encompasses longstanding grievances and unresolved assertions of rights and title. The Jasper Aboriginal Forum has enabled significant reconciliation and reconnection initiatives, helping to achieve Parks Canada's mandate for protection, visitor experiences and education – and by extension, adding cultural depth to our presentation and protection of the Canadian Rocky Mountain Parks World Heritage site. It is the right thing to do.

The healing process requires time, and encompasses longstanding grievances and unresolved assertions of rights and title. The Jasper Aboriginal Forum has enabled significant reconciliation and reconnection initiatives.

14

Legacy of a chief: Chief Roi Mata's Domain, Vanuatu

AMARESWAR GALLA[1]

The aquatic continent

The Pacific is home to extraordinary cultural and biological diversity. In September 1999 the Vanuatu National Museum and Cultural Centre organized a workshop for drafting the Strategic Plan for the Pacific Islands Museums Association. The participants, mainly directors of museums in the Pacific, stayed on for one of the most intensive and significant workshops on the World Heritage Convention convened in the region by the UNESCO World Heritage Centre. This was followed by the drafting of the Pacific 2009 Programme (2000–2009) and further developed through regional consultations and the 31st session of the World Heritage Committee Meeting held in Christchurch in 2007 under the chairmanship of Mr Te Heuheu, Paramount Chief of the Ngati Tuwharetoa Maori Tribe of New Zealand. These intensified efforts resulted in an increase in the number of Pacific States Parties to the World Heritage Convention as well as their World Heritage sites.

The Pacific Islands subregion, comprising a third of the world's water, is sometimes called the aquatic continent. Fourteen out of a potential seventeen countries in the Pacific are States Parties to the World Heritage Convention. Most of them have very small populations with over 90 per cent indigenous peoples. About 10 million people speak one-fifth of the world's languages. It is

---

[1] This study is based on material provided by Marcelin Abong, Director of the Vanuatu National Museum and Cultural Centre. Ralph Regenvanu, Former Director of the Vanuatu National Museum and Cultural Centre and until recently Chair of the Vanuatu National Cultural Council has been generous with time and support providing a range of published and unpublished materials.

The grave on Artok island with the body of Roi Mata is surrounded by over fifty of his family and retainers. It is an outstanding example of a mass single-event chiefly burial.

170

... the Pacific Islands have the highest rate of indigenous peoples within their national populations of any region in the world.

said that the Pacific Islands have the highest rate of indigenous peoples within their national populations of any region in the world, as well as the highest proportion of land that is held under customary ownership or traditional land and sea tenure systems (Regenvanu, 2008).

As a follow-up to the Pacific 2009 Programme, the States Parties prepared the Action Plan for the next five years (2010–2015) at the Pacific World Heritage Workshop held in Maupiti in 2009. They examined progress in the implementation of this plan at the Pacific World Heritage Workshop hosted by the Samoan authorities (Apia, 5–9 September 2011). The workshop brought together member states, advisory bodies and regional organizations such as PIMA (Pacific Islands Museums Association) and ICOMOS Pasifika, discussing issues specific to the Pacific such as the Pacific Heritage Hub, a facility aiming to enhance knowledge management, capacity-building and sustainable funding/partnership-building in the Pacific region in the field of heritage management.

Indigeneity and sustainability

Chief Roi Mata's Domain (CRMD) is the first World Heritage site in Vanuatu. It consists of landscapes and waterscapes delineated by three sites dating back to the early 17th century. These sites are on the islands of Efate, Lelepa and Artok. They are associated with the life and death of the last paramount chief, or Roi Mata, of what is now central Vanuatu. The sites include Roi Mata's

Intergeneration transmission is crucial for safeguarding the Outsanding Universal Value of this continuing landscape. The power of oral tradition and archaeological investigation combine to protect the landscape of Roi Mata.

residence, the site of his death and his mass burial site. The whole area is closely associated with the living traditions surrounding the chief and the moral values he espoused. The site reflects the convergence between oral traditions and archaeology. It is evidence for the continuation of Roi Mata's social reforms and conflict resolution, still relevant to the people of the region who are the responsible custodians.

Sustainable development of CRMD is ensuring the continuity of the outstanding universal value of the site and in doing so working through the primary stakeholders to find pathways for future development of the local communities so that the benefit sharing does not diminish the inheritance of the *ni*Vanuatu (indigenous peoples). The inscription criteria succinctly underline the intergenerational responsibility that is the essence of sustainable development.

Criterion (iii) Chief Roi Mata's Domain is a continuing cultural landscape reflecting the way chiefs derive their authority from previous title holders, and in particular how the tabu prohibitions on the use of Roi Mata's residence and burial site have been observed for 400 years and continue to structure the local landscape and social practices.

Criterion (v) Chief Roi Mata's Domain is an outstanding example of a landscape representative of Pacific chiefly systems and the connection between Pacific people and their environment over time reflected in respect for the tangible remains of the three key sites associated with Roi Mata, guided by the spiritual and moral legacy of his social reforms.

Criterion (vi) Chief Roi Mata's Domain still lives for many people in contemporary Vanuatu, as a source of power evident through the landscape and as an inspiration for people negotiating their lives.

Inherited responsibilities – governance

Vanuatu has developed over the past three decades one of the most systematic programmes for inventorying living heritage of over 100 language groups

Interpreting outstanding universal value by local stakeholders through experiential heritage tourism contributes to job creation for local community members. Here the audience are participants at the first Oceanic Art Symposium in May 2008, including most of the experts currently responsible for Pacific World Heritage sites.

through the first voice of the bearers and transmitters. This unique initiative, known as the Community Fieldworkers Program, is directly managed and driven by the practitioner communities in the management of their own living heritage. The Vanuatu National Museum and Cultural Centre provides the hub for this safeguarding process. The Centre also implements research policies stipulated by the Vanuatu National Cultural Council. It is further responsible for maintaining a register of national heritage sites, including those protected under the Preservation of Sites and Artefacts (Amendment) Act (2008). The Centre as a core stakeholder builds on this role, trust and knowledge base as the current responsible agency for the State Party administration of CRMD.

The World Heritage inscription through indigenous leadership and negotiation is an ongoing process, over and above the listing. The buffer zone and strategies for minimizing negative impacts on the outstanding universal value of the site are new concepts and subject to much discussion and at times contestation, involving considerations of land use planning and new forms of leasing land to outside investors.

Community activities

Inscription on the World Heritage List has become a crucial step for local economic development in many parts of the world, especially countries

Local women provide indigenous meals as part of Roi Mata Domain experiential tourism. This generates income for women who are carriers and transmitters of local flora and culinary knowledge systems.

with low economic indicators. CRMD and Vanuatu are no exception. The extent to which these expectations can be realized varies based on the location and the local and international infrastructure that is accessible to the primary stakeholder communities. Several projects are in the scoping and developmental stages. World Heritage status has both direct and indirect benefits that are not easily quantifiable in a small island country such as Vanuatu.

Brigitte Kalmary Laboukly, responsible for coordination of CRMD at the Vanuatu National Museum and Cultural Centre, raises a number of important considerations. She has identified three key benefits derived to the community from the inscription:

- Unifying influence within the Lelema community. The property management system is founded on traditional, communal systems of land tenure. The land-owning chiefs are paid an annual tribute for the use of their land by Roi Mata Cultural Tours and international visitors, bringing economic benefit to the Lelema community and to Vanuatu.
- Providing alternative livelihood opportunities to land leasing (Roi Mata Cultural Tours; World Heritage Bungalows).
- Facilitating training, education and awareness opportunities for the community (tourism operation; buffer zone planning workshops; futures planning workshops; legal clinics).

173

Craft development

It was International Women's Day, 8 March 2012, and the theme was rural women and poverty alleviation. In the morning a large gathering of women began discussions about sustainable livelihoods and the role of women in community development at Mangaliliu. Leisara Kalotiti, president of Mangaliliu Craft Committee, chaired the meeting. She is also deputy chair of the World Heritage and Tourism Committee (WHTC), which convened the meeting. A similar meeting was held on Lelepa Island in the afternoon.

The villages of Mangaliliu and Lelepa Island in the north-west region of Efate are collectively known as the Lelema communities. The meeting focused on the outcomes of a three-year New Zealand Aid-funded Craft Revitalization and Enterprise Program. In partnership with the World Heritage site, it was producing high-quality contemporary handicrafts using traditional techniques for sale to tourists. One of the activities included cataloguing some of the traditional crafts that were lost during colonial times, in the spirit of revitalization of the intangible heritage of the local communities. There was enthusiasm about creating a place to showcase the crafts and establish a shop.[2]

Cultural tours

In the morning I sat in as an observer with Chief Mormor and in the afternoon with Chief Richard Matanik, acting chair of the World Heritage and Tourism Committee. Chief Richard also generously took me on a tour of the island and to the house and grave of the late Chief Douglas Markfonulolowia Kalotiti, former chair of the WHTC for the Lelema region communities. Chief Douglas took us around the World Heritage site in 2009, when the photographs were taken. He recalled how he explained to the Committee members, especially chair Christina Cameron, the importance of the CRMD to his people. Where there was concern about the investors and leasing of land in the buffer zone, he said that the answer was very simple – all issues can be negotiated and settled in the spirit of the legacy of Chief Roi Mata.

Roi Mata Cultural Tours is a community-owned tourism operation that was developed together with the nomination file for Chief Roi Mata's Domain. Currently managed by WHTC, it is envisaged that operations will soon be taken over by an independent management body in compliance with the guidelines for safeguarding the outstanding universal value of Chief Roi Mata's Domain.

Chief Richard explained that the Roi Mata's Domain Cultural Tour promotes the World Heritage area through the concept of *nalfan* Roi Mata (in the footsteps of Roi Mata). It provides visitors experiential learning from the cultural landscape about the outstanding universal value and the legacy of Roi Mata. In keeping with the community's wish, the tour is designed to provide broad benefits across all families. This is achieved through the employment of men and women of all age groups to work as guides, cooks, craft manufacturers and actors for the tour. The landowners of the World Heritage area are paid

---

[2] I attended the meetings courtesy of Chris Delaney, Craft Adviser from New Zealand for the project.

Fels cave, site of Chief Roi Mata's death, is a massive circular chamber with inner wallls covered in ancient paintings and engravings, dating from the time of the last Roi Mata around AD 1600.

an annual tribute (*nasaotonga*), at which the community honours and thanks the landowners for committing their land for the purpose of operating the Roi Mata Cultural Tour.[3]

Bungalows

As a result of the inscription of the CRMD on the World Heritage List, tourism has increased. Visitors also want to have a more in-depth experience of the site. It is the desire of the community to develop World Heritage Bungalows providing overnight accommodation to visitors. This may generate income to a number of landowners and preclude the leasing of their land. It was mentioned in a meeting at the Vanuatu Cultural Centre that the World Heritage Bungalows will provide:

- an alternative income stream for landowners;
- a way of profiting from land without engaging in lease contracts;
- a way of capitalizing on existing hospitality skills within the community that have been honed as a result of the Roi Mata Cultural Tours;
- the ability to offer 'package tours' that could be marketed internationally, inclusive of overnight stays within the World Heritage buffer zone.

A World Heritage Bungalows planning workshop set up the Tupirou committee (six members, equal representation from Mangaliliu and Lelepa).

Some lessons for sharing

Indigenous Pacific leadership has consistently argued that the regional needs and especially those of small island developing states (SIDS) have to be addressed from a localized perspective, as reflected in the Barbados and Mauritius Declarations of these states. Several scholars have pointed out that the greatest challenge for World Heritage listing in the Pacific is negotiating

[3] Reports by Marcelin Abong.

Vanuatu National Museum and Cultural Centre, the agency responsible for Chief Roi Mata's Domain, presents an exhibition of the World Heritage site and young people celebrate oral traditions along with contemporary creativity and music.

176

land tenure and ensuring the capacity of indigenous peoples to participate in the economic benefits that accrue from World Heritage status (Ballard and Wilson, 2012; Trau, 2012).

Brigitte Kalmary Laboukly once again pointed out several factors to take into consideration during the 2011 World Heritage workshop in Samoa and during discussions at the Vanuatu National Museum and Cultural Centre in March 2012.

- Governments of SIDS have inadequate budgets to invest, whether financial or administrative, in World Heritage sites such as CRMD.
- The WHTC at CRMD relies heavily on foreign aid and foreign volunteers to support their activities, and members of the committee often cover the costs of their management obligations.
- The WHTC does not have the capacity to deal with issues that potentially threaten the values of the World Heritage area.
- The WHTC is not mandated to manage the buffer zone (which contains around 200 landowners).
- Too much dependence on the CRMD International Advisory Group, rather than national advisers, to define priorities for Chief Roi Mata's Domain (insufficient skills transfer).

Chief Richard Matanik explains that the use of traditional watercraft is promoted as an integral part of the World Heritage landscape and environmentally responsible tradition.

For the next generation

In July 2009, I was fortunate to be with Chief Douglas after the CRMD inscription. The centrality of living traditions to creativity and how this is informed through ritual, oral traditions, cuisine and intergenerational transmission were packed into an experiential heritage tour that is exemplary. The quality of what can be offered promises potential benefits from tourism.

In 2012, a new initiative will bring together people in the CRMD. Chief Richard Matanik and Leisara Kalotiti, mentioned as chair and deputy chair of the WHTC, and the community have been working on the launch of a festival in memory of the late Chief Douglas Kalotiti who appealed to the World Heritage Committee for the inscription of the site at the 2008 Quebec meeting. The landscape is one of layers of memory. The new festival is appropriate for safeguarding the site's outstanding universal value. Douglas is no longer with us. But he leaves for the next generation a tradition of leadership. Sustainability is understood by the primary stakeholder community as ensuring the continuity of this intergenerational transmission.

177

15 Living cultural landscape: Rice Terraces of the Philippine Cordilleras

JOYCELYN B. MANANGHAYA[1]

Unique and traditional land use

The Rice Terraces of the Philippine Cordilleras are a product of tradition with emphasis on the agricultural cycle of the indigenous Ifugao people and the role of cooperative systems in creating and maintaining the rice terraces and the forests. Local communities are the drivers for the conservation of the site's outstanding universal value and for securing sustainable development. The seamless and organic human interaction and adaptation to the environment, and the evolution of a unique land-use system over two millennia by the Ifugao, constitute the essence of the steep terraced landscapes. They are aesthetically appealing and testimony to a predominantly communal system of rice production through small-scale farming. Harvesting water from the forested mountains above is integrated with the management of an irrigation system of stone terraces and ponds. The challenge is to sustain the continuity of the landscapes in the face of social and economic change.

In the rice terraces of the remote areas of the Philippine Cordillera mountain range on the northern island of Luzon, the local Ifugao communities are the primary stakeholders in the five clusters of terraces that comprise the World Heritage site:

the Nagacadan terrace cluster in the municipality of Kiangan, manifested in two distinct ascending rows of terraces bisected by a river;

the Hungduan terrace cluster that dramatically emerges into the shape of a spider web;

[1] Heritage conservation architect, Philippines.

Batad, traditional settlement in present times. Members of the Batad community are now benefiting from tourism as they own accommodation facilities and act as tour guides to show visitors around.

179

the central Mayoyao cluster which is characterized by terraces interspersed with traditional farmers' houses (*bale*) and granaries (*alang*);

the Bangaan terrace cluster in the municipality of Banaue that backdrops a typical Ifugao traditional village; and

the Batad cluster of the municipality of Banaue that nestles in amphitheatre-like semi-circular terraces with a village at its base.

The Ifugao are key partners in a system of sustainable living which, among other factors, includes various considerations for sustaining territories within and adjoining the World Heritage site, management and governance issues, and the economic development of the landscapes. The rice terraces are a product of cultural traditions that encompass agricultural practices and rites believed to ensure good harvest. Traditional cooperative systems (*ubbu, danga, baddang*) guarantee the availability of help for the expansion and maintenance of the fields and the harvest of crops with communities living within it playing an important role in the agricultural practices and associated rites.

The rice terraces are a product of tradition and the role of the cooperative systems of the primary stakeholder, the Ifugao communities, is at the heart of their creation and maintenance.

The agricultural practices in the rice terraces and their accompanying rites and rituals are an integral part of the indigenous knowledge systems. The

The rice terraces are a product of tradition and the role of the cooperative systems of the primary stakeholder, the Ifugao communities, is at the heart of their creation and maintenance.

principal ones are the *Lukya* with its associated ritual, the *Ubaya*; the *Ahigabut* with its associated ritual, the *Hipngat*; the *Panal* with its associated ritual, the *Pingil*; the *Kulpi* and the ritual performed after a month – the *Hagophop* and the *Ahikagoko*; the *Ahidalu* and its accompanying ritual the *Bodad*; and when the rice plants have matured, the *Paad*; and when during the *Ani* period or harvest time – the *Ahi-ani* and the *Ahi-tul-u* and its associated rituals the *Tuldag* and the *Ahi-Bakle*. And after harvest time, the *Upin* is performed and so is the *Tungo*, after the *Ahi-Bakle*. And then there is the *Kahiw* ritual (Conklin, 1980; Dulawan, 2001).

Men and women work together in the fields throughout the agricultural cycle but there are gender-specific tasks. Men prepare the land for planting while groups of women harvest the crops. Priests are mostly men, although women with ritual knowledge intercede on rare occasions. The traditions and rituals of farming dictate the everyday life of the communities and their cooperative systems assure a good harvest. The Hudhud or narrative chants are performed by the Ifugao during the rice sowing season, at harvest time and at funeral wakes and rituals.[2]

The Ifugao are matrilineal and the wife generally takes the main part in the chants. Her brother occupies a higher position than her husband. The stories tell of ancestral heroes, customary law, religious beliefs and traditional practices, and reflect the importance of rice cultivation. The narrators, mainly elderly women, hold a key position in the community, both as historians and preachers.

Indigenous knowledge and sustainability

Sustainable development in the rice terraces provides for the essential development needs of the people and minimizes the negative impacts on the natural and cultural environment. It allows for the continuity of traditional practices and the dynamism to adapt to changes. Sustainability is closely linked to such terms as 'conservation of the outstanding universal value' of World Heritage properties but moves on to include 'regulated economic development' in consideration of the peoples' needs (Mananghaya, 2011). Equilibrium is maintained through the continual performance of traditional practices, proper utilization of the land and natural resources and careful selection of income-generating activities that are appropriate for the place.

[2] The Hudhud Chants of the Ifugao were proclaimed as a Masterpiece of the Oral and Intangible Heritage of Humanity in 2001 and inscribed on the Representative List of the Intangible Cultural Heritage of Humanity in 2008.

Ifugao has also seen an improvement in the development of cultural industries such as carving and other by-products of culture and nature.

181

As water flows from the forest-clad mountains, the communities ensure the integrity of the forests through traditional forest-management practices. These include responsible harvesting of forest products, making certain that trees that retain water for irrigating fields are left untouched while those ideal for wood carving and other income-generating purposes are not over-exploited. With forests left intact, the supply of water to irrigate the fields is equally assured.

The communities' role in retaining the integrity of the whole landscape is an important factor in ensuring sustainability of the outstanding universal value. Integrity is attained when the rice fields, waterways and irrigation canals are continually maintained by the local people. Integrity is also achieved when abandoned areas are reclaimed and cultivated. In Ifugao, fields that are to be left untilled by family members could still be maintained by relatives with an agreement to do so. The close kinship of the people helps to pass properties from generation to generation and assists in their continued use and maintenance. In instances where a family is unable to directly farm their fields, neighbours or relatives are contracted to help. The preparation of fields for planting is passed on to a neighbour while the owner provides food for the harvesters and in turn shares half of the produce.

The maintenance of the lifeline of the rice terraces, the irrigation system, is likewise sustained by the community. Regular maintenance is essential in ensuring the flow of water to the rice fields. In instances when the irrigation canals are damaged by clogs or debris, or when collapse of the terrace occurs, the owners of the fields directly affected by the damage immediately work together in a cooperative manner, declog the canal or waterway and restore order.

Integrity lies in the maintenance of the fields' original function for the planting of rice and/or its associated vegetable crops. As there are a number of traditional land uses in the terraces such as the rice field (*payo*), swidden farm (*habal*), forest (*muyong*), and village (*boble*), integrity of the landscape as a whole is sustained and regulated by the communities as well as the landowners who make the necessary decisions on land use.

Harvest festival. The local communities play an important role in agricultural practices and associated rites.

The Ifugao people are particularly aware that conversion of rice fields for other uses affects the quantity and quality of the rice harvest, which in turn impinges on family food supplies. Similarly, the increase in size and number of swidden farms, although increasing the family income when planted with vegetables, considered as cash crops, unfortunately leads to the loss of forests, which in turn may affect the water supply to the fields.

The integrity of the landscape is affected when villages expand to the extent that they encroach on forests. The maintenance of villages in their original locations, while allowing for regulated sprawl in village size to accommodate a reasonable increase in population, ensures that the rice fields and forests are kept intact. Similarly, the proper use of forests that avoids the slash-and-burn (*kaingin*) system for the creation of new swidden farms prevents deforestation and ensures the retention of tree species important in watershed management. All these actions are stabilized and ably managed by the owners of these ancestral lands and as a group, by the communities themselves who live within the confines of the terraced landscape.

From traditional cooperatives to contemporary organizations

The rice terraces were declared National Treasures in Presidential Decrees 260:1973 and 1505:1978. The Ifugao Terraces Commission with presidential mandate for the preservation of the rice terraces was established in February 1994. The terraces are protected by the Republic Act No. 10066:2010 on the conservation of the National Cultural Heritage. The site is under the management of the Provincial Government of Ifugao and the National Commission for Culture and the Arts. A Rice Terraces Master Plan comprehensively covers management, conservation and socio-economic issues.

Ifugao traditional dance. The Hudhud and narrative chants are performed by the Ifugao during the rice agricultural season, at harvest time and at funeral wakes and rituals. Women generally take the main parts in the chants.

The World Heritage site clusters of terraces are administratively co-managed by local government units and national government agencies, which were identified as site managers when the property was listed in 1995. However, the rice fields, the land within the villages and the houses, the woodlots (*muyongs*) and the swidden farms are privately owned by local people, a number of whom live within the terrace clusters. Management of the rice terraces is carried out collaboratively by these entities, with each playing a specific role in maintaining the fields and surrounding environment. Within a balanced system that allows for sustainable development, they help each other in supporting the social, economic and environmental pillars of sustainable development.

183

The local government has launched programmes to ensure the continuing rehabilitation of areas within the World Heritage site that have been affected by typhoons and natural disasters. These programmes include support to local communities so that major damage that could not be resolved through the *ubbu* system because of its significant size, scale or gravity could immediately be restored so as not to disrupt the agricultural cycle. Support is given in the form of funds, food or other services in exchange for labour. The communities contribute by providing the necessary materials for the work, including their labour.

NGOs such as the Save the Ifugao Terraces Movement (SITMo) have been instrumental in carrying out projects that equally benefit the rice terraces and the people. NGOs facilitate tourism programmes and technical projects such as drawing up land-use plans.

Officially instituted cooperative systems encouraged by government also emerge as a vital force that ensures constant utilization of the fields. In line with tradition, the owners of the rice paddies come together to help in the preparation of the fields prior to sowing and planting. It is the same group who ensure the successful harvesting of crops. To date, the former traditional cooperative systems have been reconstituted into what are called Rice Terraces Owners (RTO) organizations. There are around 774 RTO organizations throughout Ifugao province registered with the Department of Labor and Employment (DOLE) as of December 2011 (Ifugao Cultural Heritage Office, 2011), which

Ifugao workers rehabilitating a collapsed area. The local government has launched programmes to ensure the rehabilitation of areas affected by typhoons or natural disasters.

184

help government in the rehabilitation of damaged areas as well as in the harvesting and selling of crops.

Similarly, NGOs such as the Save the Ifugao Terraces Movement (SITMo) have been instrumental in carrying out projects that equally benefit the rice terraces and the people. NGOs facilitate tourism programmes and technical projects such as drawing up land-use plans.

Harmonization of outstanding universal value and community ownership

The realization of stakeholder community benefits has become part of sustainable development as a framework for progress for this World Heritage site. A significant decline in the performance of traditional practices has been reversed with proactive programming by government in the revitalization of local heritage and outstanding universal value. The yearly fiesta (*Gotad*) celebrated by the communities demonstrates the Ifugao's determination to continue cultural enhancement and development. Teaching children an understanding of the Hudhud chants also permeates the importance of this oral tradition into the subconscious of the young – critical for intergenerational transmission of heritage values and outstanding universal value.

For 2,000 years, the high rice fields of the Philippine Cordilleras have followed the contours of the mountains.

Documentation of heritage with the help of community elders and the transmission of traditional skills to children through Ifugao State University's Nurturing Indigenous Knowledge Experts (NIKE, 2009) programme, now in its fourth phase of implementation, allow for sustainable rice terrace conservation.

Similarly, there have been benefits in promoting traditional products in the rice terraces. There are terraced rice fields in many parts of Asia, but few are at a higher altitude or on steeper slopes than the Philippine terraces. High-altitude cultivation requires a special strain of rice that germinates under freezing conditions. It grows chest-high with non-shattering panicles and can be harvested on slopes that are too steep to permit the use of animals or machinery of any kind.

The high nutritive quality of the traditional *tinawon* rice, the only variety associated with traditions and rituals, makes it three to four times more valuable than the regular rice from the lowlands. The *tinawon's* very high quality compensates for its smaller quantity, being harvested only once a year. To date, traditional rice from the Philippine Cordilleras is being exported to the lowlands and to countries such as the United States.

Incentives and benefits realized from growing this traditional variety of rice have in turn resulted in an increase in the number of RTO organizations and members. This increase is also a result of the encouragement extended by government through its support for the restoration of collapsed parts of the rice terraces. As mentioned above, it is the RTO community members who undertake the repair work, with government providing financing or support in kind for the contract.

Ifugao has also seen an improvement in tourism and the development of cultural industries such as weaving, carving, rice wine-making and other by-products of culture and nature. These traditions have attracted tourists wanting to learn and know more of the local culture. The Ifugao rice terraces are considered to draw in almost as many tourists, both local and foreign, as Baguio City and Sagada, also in the Philippine Cordilleras. Members of the

185

Rice Terraces, Ifugao Province. The rice terraces are a dramatic testimony to a community's sustainable and primarily communal system of rice production.

community of Batad for example, are now benefiting from tourism as they own accommodation facilities and act as tour guides to show visitors around. To date, the negative impacts of tourism have not yet greatly affected the rice terraces, although a tendency towards an unregulated flow can be seen in some areas due to the beauty of the site. This has resulted in a clustering of accommodation facilities affecting formerly traditional intact landscapes. As there is a great desire to generate income from cultural and community-based tourism, close monitoring of developments that could lead to negative impacts is needed to prevent disaster.

Following encouragement by local government to improve the production of culture related by-products, an increase in products that promote the heritage mark of Ifugao has also been felt, further encouraging and supporting community cooperation. A specific example is the famous Ifugao *tinawon* rice, the marketing of which has directly profited the communities. Rice wine bearing Ifugao's mark, traditionally grown coffee and other goods have also proved attractive to the lowland market, with communities being the direct producers again benefiting from the sales.

Lastly, the recent publication of maps, through participatory community engagement, that define the core and buffer areas of the five clusters of terraces in the Ifugao World Heritage site will help to ensure the sustainability of the outstanding universal value. Defining the conservation areas is perceived to be both beneficial and challenging to the communities living within the site. It allows for the formulation of a regulated development policy which is thought to support communities' and people's rights to make decisions on the use of their land. Boundary delineation will help to address overdevelopment or uncontrolled development within the World Heritage site and in other areas. In the end, the communities will benefit from a regulated policy where they themselves partake in decision-making both as owners and beneficiaries of a sustainable environment.

Women in field. There are gender-specific tasks in the fields. Men prepare the land for planting while groups of women harvest the crops.

Local actors in World Heritage conservation

In the Rice Terraces of the Philippine Cordilleras, the safeguarding of the site's outstanding universal value is assured through a localized framework of sustainable development, understood by the management and the Ifugao as 'economic growth, social development and environmental protection in order that future generations have the same or better opportunities than the present generations' (Tolba, 2004). The understanding and practices are consistent with the recommendations of the Brundtland Commission and Local Agenda 21. The equilibrium of communities, environment and economy ensures the continuity of the outstanding universal value. Collaborative partnerships between government, private sector and communities are important in ensuring sustainability. The regular updating of management and conservation plans has become important in the achievement of optimum benefits for the local communities through economically productive but regulated activities.

The impacts of tourism are not yet evident at the World Heritage rice terraces. A Tourism Master Plan is vital to prepare for the benefits and impacts of tourism. Similarly, as land-use plans are still being drawn up, the clear definition of all the site's attributes – cultural and natural, tangible and intangible – is important to give a better understanding of its outstanding universal value.

The local communities living within a World Heritage site can play an important role in achieving sustainable development. The Ifugao constitute the civic agency instrumental in ensuring regulated economic progress, the lynchpin guaranteeing sustainability of the territories within and adjoining the rice terrace clusters, the culture and traditions, and the natural environment in its political, social and economic context.

187

16

The strength of a cultural system: Cliff of Bandiagara (Land of the Dogons), Mali

LASSANA CISSÉ[1]

One of West Africa's most impressive sites

The Cliff of Bandiagara (Land of the Dogons) is one of four Malian sites on the World Heritage List. It was inscribed as a mixed site (cultural and natural) in December 1989, under criteria (v)[2] and (vii)[3]. Mali is a vast continental country in West Africa, covering 1,204,000 km[2] and with a population of 14 million. The site in the Mopti region is quite large, one of the largest World Heritage mixed sites, extending over 4,000 km[2]. Over two-thirds of the inscribed area is covered by the Bandiagara plateau and cliffs, over 100 km south-west to northeast. This area includes 289 villages located in three natural regions: plateau, cliffs and plains. The populations settled here are mainly classified as Dogon, hence the additional designation of the site as Land of the Dogons.

The sustainable and productive management of World Heritage sites in Africa and the strict adherence to the principles of the 1972 World Heritage Convention require a better understanding, perception and interpretation of the concept of World Heritage on the part of population groups living in and around heritage sites. In working towards these aims the following questions have provided guidance:

[1] Site manager, head of the Cultural Mission of Bandiagara, Mali.
[2] Outstanding example of a traditional human settlement, land-use, or sea-use which is representative of a culture (or cultures), or human interaction with the environment especially when it has become vulnerable under the impact of irreversible change.
[3] Contain superlative natural phenomena or areas of exceptional natural beauty and aesthetic importance.

■ How can communities be involved in heritage management programmes and projects, in the context of local development?

■ How can World Heritage status be of benefit to local populations and engage them in sustainable conservation of heritage sites and their resources?

■ What kind of national legislation needs to be drawn up, in tune with certain local conventions (those of local communities), based on an endogenous definition of heritage, for rational and sustainable site management?

These questions were the background to the shared vision necessary to consider the management of Land of the Dogons World Heritage site with the dual aims of:

■ permanently preserving the tangible and intangible values linked to both cultural and natural heritage within the site;

■ ensuring the participation of local populations in the development of activities on the site, improving their living conditions, and contributing to local development.

Cultural action in community engagement

The Cultural Mission of Bandiagara, with support from the World Heritage Centre and in collaboration with the German University of Konstanz, conducted an inventory and documentation of traditional Dogon architecture in 1996 and 1997. This work led to the implementation in 1999 of the Ecotourism in Dogon Country project, designed in cooperation with the communities living on the site and the German Development Service (DED, currently GIZ) with the following objectives:

■ Improve the living conditions of the rural populations living on the site, through participatory management of local cultural and tourism resources.

■ Preserve and promote the cultural and natural heritage site of Dogon Country.

■ Minimize the negative impact of tourism on the sites.

■ Strengthen the self-promotion of rural populations and their deeper involvement in the exploitation of heritage resources as a means to increase their income.

■ Promote new tour routes in other cultural and natural areas with potential, to better control the load capacity of sites.

In the past twelve years, a number of projects have been implemented. These include the construction of cultural infrastructure (village museums) and

Construction in local material, Bandiagara. In the past twelve years cultural and tourism infrastructures have been implemented at the Bandiagara site.

tourism structures (community settlements). Cultural and natural sites have also been developed with actions geared toward the promotion of traditional crafts and safeguarding intangible heritage.

The following are examples of some of the specific actions undertaken:

- Construction of a museum in Nombori: this museum is managed by the village community which is developing other collective actions to make tourist activities profitable (local crafts, federated canteen management in the museum and marketing local products).
- Rehabilitation of a community camp at Kani Kombolé.
- Support for a group of women dyers to promote local textile skills and crafts.
- Construction of a museum of arts and crafts at Enndé.

These achievements are the result of local initiatives supported by the Cultural Mission, in which the beneficiaries – local communities – are also the drivers of sustainable development.

The Ecotourism in Dogon Country project has improved understanding that the sustainable management of the World Heritage site must be facilitated in a collaborative framework that brings together the tourism and heritage sectors, along with the local communities. All these elements must be present and play an active role from the outset. Tourism relies on the cultural and natural heritage values of the Dogon people but these must be maintained and transmitted to future generations; this cardinal mission belongs first and foremost to local communities living on the site, for they are the custodians and guardians of their heritage.

One of the first priorities was the development of a management and conservation plan for the site (2006–2010). It included the participation of all stakeholders within the inscribed perimeter. This plan, funded in 2005 by the World Monuments Fund, and developed with the technical and methodological support of CRAterre,[4] is currently being updated for the next five years (2012–2016).

[4] Center for the Research and Application of Earth Architecture, part of the School of Architecture of Grenoble (France).

As an integral part of the economic sustainability of the site, a capacity-building programme was implemented for the local stakeholders. Workshops on the economic value of heritage were brought to the attention of elected officials and representatives of the fields of culture and tourism.

Cultural infrastructure was developed through the establishment of village museums and art and crafts centres, with the support of UNESCO, the European Union and the World Bank, in the villages of Songo, Soroly, Bandiagara and Sangha. A separate centre was constructed for craftsmen in Koundou Guina.

The restoration, rehabilitation and development of ancient sites and monuments were carried out in the cliffs area and on the Dogon plateau, in the villages of Téli, Kani Kombolé, temple d'Arou, Néni and Banani, supported by the Netherlands Development Cooperation, the World Monuments Fund, and implemented with the technical collaboration of an architectural firm in Mali (AUDEX) and CRAterre.

A significant project focused on the improvement of the water supply to high-altitude settlements, together with the implementation of a solar pump system in the village of Bolimba.

Another project on the valorization of local architecture and materials, strengthening local capacities for a better contribution of the construction sector to the sustainable development of Dogon Country, funded by the European Union, was conducted in cooperation with CRAterre, the NGO RADEV-Mali, and with the financial support of Misereor and the Abbé Pierre Foundation.

The development of walking trails in the cliffs was part of the restoration programme funded by the World Monuments Fund as part of the implementation of the Management and Conservation Plan.

The appropriation of heritage management and local development must necessarily include the identification, planning and implementation of endogenous programmes. It is important to understand that the sustainable management of a living site such as Dogon Country should encourage community participation and promote the local economy.

Furthermore, development brings change, which involves people who must become central in all processes dealing with change. Local communities have their needs and priorities; socio-economic and cultural activities should mainly relate to their implementation.

It is important, for the conservation of the outstanding universal value of a site and for the harmonization of local development programmes, to develop strategies in which dynamic conservation and income-generating activities overlap and become complementary.

191

It is important to understand that the sustainable management of a living site such as Dogon Country should encourage community participation and promote the local economy.

A blacksmith and woodcarver making a *Kanaga* mask with his son in Ogol du Haut village.

Decentralization and community development

In 1991, Mali began the long process of political democratization and initiated a process of decentralization. This culminated in 1999 with the creation of 703 districts throughout the territory. Decentralization aims, among other factors, towards the greater empowerment of communities that need to ensure their own development, based fundamentally on the exploitation and development of local resources. Each district created was required to plan and implement a five-year development programme, based on the socio-economic and cultural realities within the local territory.

On the Bandiagara plateau, twenty-one districts were created and are run by councils directed by elected mayors. Some district authorities have dreamed of turning the exploitation of the cultural values of the site into the core of local development through cultural tourism. Attempts to develop local tourism have included the implementation of tolls to collect taxes to access the sites, municipal taxes, the organization of paying cultural and artistic events during the high tourist season, and taxes on other activities related to heritage resources.

In some locations, heritage management and tourism in the context of decentralization in Dogon Country has created a chaotic development of private initiatives, with the proliferation of reception facilities that are inadequate for the local built environment and landscape. The site management has been focusing on minimizing the negative impacts on the outstanding universal value of the site caused by some of the tourist attractions.

On the other hand, there is a renewed interest in some communities concerning the revitalization of their heritage and its exploitation for local development: many cultural associations have been formed to encourage responsible tourism and the preservation of cultural heritage: 'the recent evolution of cultural events in two of the most touristic Dogon locations give us a glimpse into

Head of the village of Pelou, situated on the Bandiagara plateau. The Bandiagara plateau and cliffs, of over 100 km south-west to north-east, are the site of 289 villages.

193

the links that can gradually develop between the decentralization process, its difficult implementation, and the evolution of contemporary traditions. In this highly recognized region, that is also popular for its traditional "customs", the introduction of districts has given rise to new manifestations and activities that, while being aimed primarily at exogenous participants, make sense to local stakeholders' (Doquet, 2006).

Several recent local businesses result from the new relationships that are being developed among populations, villages and visitors interested in the development of Dogon Country.

Although still in its early stages, local development bodes well for the ownership and administration by local stakeholders and communities of heritage and tourism. There is still a need to support local initiatives that must fit into a plan for the conservation and management of the World Heritage site.

Family dwellings. The Cliff of Bandiagara is an outstanding example of a traditional human settlement, representative of the Dogon culture.

194

World Heritage in social and economic development

To promote a local economy around the management of cultural and natural heritage resources, the following strategies should be considered:

- Develop an endogenous strategy for the development of tourism on sites for the benefit of local communities, decentralized communities and private entities.
- Reinforce the capacities of local operators in the framework of the cost-effective management of heritage resources.
- Develop adequate visitor and accommodation facilities (create an environment to enliven certain monuments).
- Create a synergistic framework for cultural resources (architecture, arts and crafts, artistic intangible heritage) and their integration with local development programmes: creation, organization and equipment of adapted structures, art and crafts centres.

Local children waving. Through the World Heritage inscription, a revitalization plan was prepared with a stress on information and education.

195

- Organize visits around heritage elements and events (historic places, cultural sites and monuments, dance and music entertainment and festivals).
- Develop micro-credit grants through a system of 'cultural banks' (loans guaranteed through the storage of art objects of value).

Participatory management in conservation and sustainable management

The local development projects initiated by various partners are all opportunities for properly aligning economic and social interests with cultural interests, and this includes the conservation of the site. Conversely, poorly planned projects can prove destructive. To ensure a positive outcome, a participatory and dynamic heritage management policy requires:

- Capacity-building among communities, local representatives and local authorities within territorial collectivities.
- Identification, participatory inventory, documentation and promotion of local building cultures as regards infrastructure, economic and social development (public buildings, houses and shops).

Local women grinding grain.

- Upgrading and revitalization of heritage-related occupations (guilds of masons, castes of craftsmen and traditional construction workers).
- Official promotion and preservation of traditional knowledge to boost the local economy through the development of cultural and natural resources: local products, corporations of masons, groups of artisans, local cultural associations.

Thus all possible means should be employed to develop a comprehensive strategy for managing cultural, spatial, environmental and architectural dynamics:

- Control the development and anarchic occupation of inhabited areas and agricultural land within the perimeters of the inscribed sites.
- Improve the living environment and housing conditions by hygiene and sanitation measures, managed at local level.
- Maintain the integrity of sites, monuments and other still functional places of worship by making specific adjustments respectful of the cultural spaces built to maintain the site's outstanding universal value.

Funeral ritual where the Kanaga masks pay tribute to the departed. The region is one of the main centres for the Dogon culture, rich in ancient traditions and rituals, art and folklore.

Revitalization through World Heritage inscription

To implement the 1972 World Heritage Convention, the Malian Government created the Cultural Missions as decentralized management institutions responsible for assisting the Ministry of Culture in its task to protect and valorize all national sites inscribed on the World Heritage List. The Cultural Missions of Bandiagara, Djenné and Timbuktu were set up by Decree 93-203 P-RM of 11 June 1993. They deal with standards in conservation, research, information and public awareness as well as the promotion and valorization of heritage resources in a context of local development. This new experience in Mali has contributed to a renewed interest in local heritage management, which systematically involves communities living on the sites in all activities.

The experience of eighteen years in the management of this vast conservation area allowed us to understand that managing a living site requires a strong involvement on the part of local communities. This participation cannot be effective unless it takes into account all matters related to subsistence and development.

Taking advantage of the decentralization policy in Mali, the involvement of communities, local elected officials, religious leaders and traditional leaders in the participatory management of heritage resources, including tourism, may take place. The principles of sustainable conservation of heritage and the challenge of development are at stake.

In this context, a number of guidelines can be set out to help answer the following question: how to preserve a living site, where people are confronted daily with subsistence and/or survival problems?

First, it is important to develop a management and conservation plan, with the participation of all stakeholders, carefully considering its implementation within the context, and including:

■ a sector plan for information, education and communication aimed at local stakeholders and the public (visitors, different social strata …);
■ a programme of research and conservation of the tangible and intangible elements that constitute the local and universal values of the site;

- a promotional strategy to reinforce the various components of local heritage, their location and interest;
- planning of synergistic development actions that incorporate heritage resources.

Maximum involvement, through specific actions, is crucial to the resolution of basic concerns, such as:

- access to safe water and energy sources;
- hygiene, sanitation and environmental issues;
- job creation at the local level to minimize rural exodus due to consecutive droughts, through the implementation of cultural and tourism infrastructures;
- site development for the profitable management of an expanding cultural tourism, evolving into true solidarity tourism, for the benefit of local communities and local development stakeholders;
- a significant contribution to reducing the extreme poverty of those living on the sites, through the support of local initiatives and management efforts, and the economic exploitation of natural and cultural heritage resources;
- implementing information centres and visitor facilities for tourists, that help to support local people in the improvement of site visiting conditions and accommodation offers.

4

Living Heritage and Safeguarding Outstanding Universal Value

The decision of the World Heritage Committee in 1994 to take into consideration the principles and views contained in the Nara Document on Authenticity (1994) in its evaluation of properties nominated for inclusion on the World Heritage List is a turning point in the history of the World Heritage Convention. It generated an enriched World Heritage discourse and listing of sites that demonstrated both cultural diversity and heritage diversity. Significantly, the knowledge of community groups living in World Heritage sites has become important in management, and this was further underscored in 2007 by the Committee adopting 'Communities' as one of the 5 'Cs', or Strategic Objectives for facilitating the implementation of the Convention.

This chapter presents five case studies that illustrate the participation of local communities living in and around World Heritage sites and contributing to the safeguarding of their respective outstanding universal value.

iSimangaliso Wetland Park (South Africa) clearly demonstrates that conservation of a World Heritage site in partnership with the primary stakeholder community can result in economic, social and environmental benefits derived to communities that have been historically disadvantaged. Conservation and community development are facilitated as sustainable development of the World Heritage site.

Sian Ka'an (Mexico) is an example of participatory methodologies and project-based learning in safeguarding outstanding universal value. It recognizes that the high degree of biodiversity conserved in the World Heritage site is partly a legacy of the traditional knowledge systems of the Maya people. It respects and benefits from the Maya management practices and landscape skills over the centuries. In doing so the approach stems the decline of traditional knowledge.

In the Republic of Korea's Hahoe Historic Village, the recognition and knowledge of the local communities has become significant for conservation and in facilitating cultural experiences for visitors. World Heritage status has also helped Hahoe villagers in their struggle to resist external appropriation of their culture and to reclaim stewardship of their village, leading to tangible economic and social benefits.

Kaiping Diaolou and Villages (China) World Heritage site presents a relatively recent phenomenon where the safeguarding of World Heritage is a networked exercise with the international diaspora. However, local people who live within the site take on shared responsibility and custodianship of its outstanding universal value.

The last case study in this chapter is the Shiretoko World Heritage site (Japan), which argues that the co-management of fisheries with the fishing communities yields significant benefits for conservation of the World Heritage site and for the local stakeholders. Building consensus with the fishing communities serves the common purpose of conservation and responsible economic development based on systematic monitoring of impacts.

17

Aligning national priorities and World Heritage conservation: iSimangaliso Wetland Park, South Africa

DIANNE SCOTT, BRONWYN JAMES AND NEROSHA GOVENDER[1]

Poverty among a wealth of natural resources

South Africa became a State Party to the World Heritage Convention in 1997. It was the first engagement with international heritage law after the dawn of democracy in 1994. In 1999, dedicated legislation was adopted, the World Heritage Convention Act (Act 49 of 1999), which incorporates the principles and values of the World Heritage Convention into South African law.

The Act brings a South African perspective to the management of World Heritage sites by acknowledging the urgent national need for development and poverty alleviation. It requires the government to find effective ways of combining the conservation of South Africa's unique endowment of natural resources with job-creating sustainable economic development (Porter et al., 2003). This integration of conservation and development in South African environmental legislation is unique to the World Heritage Convention Act[2], and makes iSimangaliso Wetland Park a 'new model in protected area management'.[3]

iSimangaliso Wetland Park was listed as South Africa's first World Heritage site for its outstanding universal value under three criteria of the ten recognized by the Convention. This case study shows that in the context of South Africa, as a developing country, it is crucial to both conserve the site's values and address the high levels of poverty and inequality.

[1] Dianne Scott, University of KwaZulu-Natal; Bronwyn James, iSimangaliso Wetland Park Authority; Nerosha Govender, iSimangaliso Wetland Park Authority, South Africa.
[2] Andrew Zaloumis, personal communication, (28 February 2012).
[3] This text draws on the collection of information and materials in *About iSimangaliso* (iSimangaliso Wetland Park Authority, 2012) and Cooper, 2012.

On 1 December 1999, when iSimangaliso[4] was listed as a World Heritage site, the then Deputy President Jacob Zuma noted that the wealth of natural assets of the park co-existed with terrible poverty, and that the region's challenge was 'to use its natural wealth to bring reconstruction and development'. The iSimangaliso Wetland Park Authority was mandated to secure the area's World Heritage values while creating a 'People's Park' and acting as a key macro-economic driver within government's regional development plan. In view of the regional context and the park mandate, this discussion focuses on the integration of poor communities living in and around iSimangaliso.[5]

The evaluation of iSimangaliso for World Heritage status noted that a number of issues needed to be resolved in order to protect its integrity. These included 'protection of catchment areas; locating the park within its regional development context; resolving the management structure; settling land claims; enabling resource harvesting; dealing with local community issues; restoration of degraded habitats (exotic species including plantation forests and management of St Lucia estuary); and amending boundaries' (iSimangaliso Wetland Park Authority, 2011). iSimangaliso's World Heritage status has enabled strategic focus on the resolution of many of the issues identified because it has been aligned with government priorities. Here we present the practices undertaken in iSimangaliso that integrate conservation with rural development as the basis for providing a set of 'lessons' to share with other World Heritage sites in similar contexts.

> In 1999 Jacob Zuma noted that the wealth of natural assets of the park co-existed with terrible poverty, and that the region's challenge was 'to use its natural wealth to bring reconstruction and development'.

204

iSimangaliso's outstanding universal value

The South African World Heritage Convention Act saw the proclamation of iSimangaliso in 1999 when sixteen contiguous pieces of land, covering 358,534 ha,[6] were consolidated into a single protected area. From the Mozambique border in the north to Maphelane in the south, the park is 190 km long. The Indian Ocean forms the park's eastern boundary which includes the marine

---

[4] At this time the park was known as the Greater St Lucia Wetland Park and after an intensive consultation process it was renamed the iSimangaliso Wetland Park in November 2007. This change was adopted by the UNESCO World Heritage Committee at its 32nd session on 6 July 2008. The word *iSimangaliso* means 'land of miracles' in Zulu and has a rich history dating back to the times of the Zulu Kingdom and King Shaka.

[5] iSimangaliso Wetland Park has a diverse group of stakeholders and interest groups that include tourists, recreational resource users (such as fishermen), neighbouring communities, scientists, NGOs, international conservation and heritage agencies.

[6] This figure includes the 14,200 ha of state land farmed by SiyaQhubeka Forests (Pty) Ltd that has been incorporated into the park through a Buffer Zone Incorporation Agreement.

Students from the 2011 iSimangaliso Higher Education Bursary and Support programme.

area along the coast which is about 5.5 km wide and runs parallel to the coast for the length of the park. The western boundary ranges from between 1 km and 55 km from the coast, narrowing towards the coast in the north and south.

iSimangaliso Wetland Park encompasses five major ecosystems and has a number of notable and varied land forms. Over geological time, diverse landscapes have been created giving rise to the contemporary iSimangaliso wetland system. The Lubombo mountains, in the west, and the spectacular dune systems along a largely unspoilt coastline in the east, enclose the lake systems (IUCN, 1999)[7]. In addition to its World Heritage values, iSimangaliso contains four wetlands recognized under the Ramsar Convention. The park was inscribed for its outstanding universal value under the following natural criteria:

Criterion (vii) *Superlative natural phenomena and scenic beauty.* iSimangaliso is geographically diverse and contains superlative scenic vistas along its 190-km coast, including natural phenomena and areas of exceptional natural beauty and aesthetic importance.

Criterion (ix) *Ecological processes.* iSimangaliso is an outstanding example representing significant on-going ecological and biological processes in the evolution and development of terrestrial, fresh water, coastal and marine ecosystems, and communities of plants and animals to create five ecosystems.

Criterion (x) *Biological diversity and threatened species.* The five interlinked ecosystems provide habitat for a significant diversity of African biota, including a large number of rare, threatened and/or endemic species (IUCN, 1999).

Socio-economic context of iSimangaliso

In the context of South Africa, iSimangaliso is a natural resource with ecological and functional integrity that is helping to drive the economic development

205

---

[7] There are two estuarine-linked lakes (St Lucia and Kosi) and four large freshwater lakes (Sibaya, Ngobezeleni, Bhangazi North and Bhangazi South).

Zulu woman at Veyane village hosts visitors. Considerable investment in tourism and conservation infrastructure has been made and the economic impact has been significant, particularly for women, who comprise 60 per cent of the workers employed.

and build the resilience of a region that was systematically underdeveloped during the apartheid era (Berkes et al., 1998; Jokilehto, 2006)[8].

The park is situated in the uMhanyakhude District Municipality, one of the poorest and most underdeveloped district municipalities in South Africa.[9] Over 80 per cent of households live below the poverty line and only about 13 per cent of the economically active population are formally employed. As a result, many households rely on natural resources for survival, setting up a tension between the long-term need to conserve these resources and short-term activities that lead to unsustainable use. Thus, the need for development is established so that the conservation of iSimangaliso can be achieved.

Most of the park is surrounded by communal land which was formerly part of the apartheid 'homeland' system. Land was alienated from black South

[8] Although iSimangaliso was not listed for any cultural criteria, these aspects are important for the site. There is a large repository of Stone Age and Iron Age sites which provide significant evidence of human settlement over thousands of years and rich cultural traditions (Klopper, 1992).

[9] According to the last published census undertaken in 2001, 573,331 people lived in uMkhanyakude District Council.

Africans, and through forced removals they were relocated to ethnic 'home-lands' (Platzky and Walker, 1985). After the 1994 elections, a process of land reform commenced and claimants were invited to submit land claims (Walker, 2008). The people of the iSimangaliso area lodged fourteen land claims that covered the entire park, of which nine have been settled by the Department of Land Affairs. Importantly, the framework for settling land claims in iSimangaliso (and Protected Areas) supports the conservation of the World Heritage site in perpetuity (iSimangaliso Wetland Park Authority, 2011).[10] By 2012, eight Community Trusts, covering 80 per cent of the land area of the park, had signed co-management agreements with iSimangaliso that spell out the benefits, responsibilities and institutional arrangements for co-management.

Management structure and governance

iSimangaliso Wetland Park Authority is the dedicated, statutory body responsible for managing the site through its Board and executive staff. It is accountable to the National Minister of Environmental Affairs.[11] The Authority is structured along 'business rather than bureaucratic lines, characterized by

[10] The framework for settling claims in iSimangaliso Wetland Park is in line with a National Cabinet decision in 2002 regarding the settlement of restitution claims in protected areas, World Heritage sites and state forests. In summary, this framework makes provision for the following:
   a) Land within a protected area can be owned by claimants without physical occupation through the transfer of title with registered notarial deed restrictions.
   b) Continued proclamation of the land for conservation purposes, where the land is used and maintained solely for the purposes of conservation and associated commercial and community activities.
   c) Continued management of the land as part of the national conservation estate by the responsible state conservation agency according to IUCN principles and the requirements of legislation and approved management plans.
   d) Land to remain part of an open ecological system and managed as an integrated part of the protected area of which it formed part before restitution.
   e) Loss of beneficial occupation is compensated for, through remuneration and provision of a package of benefits from the iSimangaliso Wetland Park that includes revenue sharing, mandatory partner status in tourism developments, access to natural resources, cultural heritage access, education and capacity building, and jobs through land care and infrastructure programmes.
   f) Sustainable partnerships between claimants and managers of protected areas must be established in a way that facilitates effective biodiversity conservation of the area, including economic viability. These co-management arrangements should enable parks to be managed effectively and efficiently by the state and remain unencumbered by several joint management committees and unwieldy co-management arrangements (iSimangaliso Wetland Park Authority, 2011).
[11] Regulations promulgated under the World Heritage Convention Act, 1999 (Act 49 of 1999) established the Park and the iSimangaliso Authority. These are Government Notice 4477 of 24 November 2000 and Government Notice R. 1193 of 24 November 2000.

Zulu at Veyane village demonstrate dance and music. Sculptors, painters, and dance and drama groups, through the art and drama programmes, have received support to explore the links between art and conservation.

a small, experienced and specialized management team' (Zaloumis, 2005). Different interests are represented on a Board comprising six to nine skilled and representative members, including those of the 'community' interests of land claimants, traditional authorities and local government.

Agreements with all levels of government (national, provincial and local), such as the contractual relationship with Ezemvelo KZN Wildlife, the provincial conservation agency responsible for the day-to-day conservation management of iSimangaliso, are also in place (iSimangaliso Wetland Park Authority, 2011).

In order to facilitate regional integration, the iSimangaliso Authority participates in municipal planning processes, such as the development of Integrated Development Plans (IDPs) and tourism initiatives. The management of the park's buffer zone, established to minimize the negative impacts on the World Heritage site, is another area of cooperative governance that involves the municipalities as well as various national and provincial organs of the state (iSimangaliso Wetland Park Authority, 2011).

The Authority also engages the private sector in the development and provision of tourism products and services in the park through public-private partnerships. There are currently fifty-eight tourism-activity licence holders

The huge numbers of waterfowl and large breeding colonies of pelicans, storks, herons and terns add life to the wild natural landscape of the site. Here, a giant heron at St Lucia lagoon.

and three private-sector accommodation facilities in the park (iSimangaliso Wetland Park Authority, 2010).

Co-management with land claimants is managed through agreements that iSimangaliso has signed with land claimants once a claim is settled.[12] Various stakeholder participation activities aimed at receiving input and disseminating information are also implemented.[13] To meet the many and varied challenges, the iSimangaliso Authority has developed a five-year Integrated Management Plan (IMP) that guides the ongoing protection and development of iSimangaliso, ensuring its integrity and the values for which it was inscribed on the World Heritage List. Local Area Plans (LAPs), which are subsidiary to the IMP, are in the process of being developed with land claimants through a participatory process.[14]

[12] Quarterly meetings are held with representatives from land claims trusts; and dedicated development facilitation staff members have regular meetings with traditional leadership, land claimants and other resource-use groups, such as cattle owners and *iNcema* (matting rush) collectors.

[13] These range from small stakeholder meetings, issue-based public meetings, workshops, electronic newsletters, local radio presentations, engagement with the media and a park newspaper. In addition, the Authority adheres to the public participation requirements of various legal processes, for example, environmental and planning authorization processes.

[14] LAPs provide the framework for sustainable local economic development for specific areas within the park, with due consideration given to its World Heritage status, basic human needs and constitutional rights as well as the relevant legal, social, environmental, institutional, cultural, economic and financial parameters.

Workers from local communities help to eliminate noxious plants from areas that were once used to grow timber. The land care and infrastructure development programmes have employed 427 community-based contractors.

Developing to conserve: community benefits

210

The iSimangaliso Authority has made significant progress in the conservation and development of the site; and particularly in resolving the issues identified during iSimangaliso's nomination as a World Heritage site.

The Authority's initial focus was to consolidate the park[15] and contributing to these efforts land care and rehabilitation programmes have removed 12,000 ha of commercial timber, which has improved ecosystem functioning; rehabilitated 45,000 ha through an alien plant control programme; created new habitats and provided employment for thousands of park neighbours. As part of the redevelopment process, considerable investment in tourism and conservation infrastructure has been made. For example, the land care and infrastructure development programmes have employed 427 community-based contractors creating about 45,799 temporary jobs over the last eleven years (iSimangaliso Wetland Park Authority, 2010).

The social and economic impact of contract work for the iSimangaliso land care programme has been significant, particularly for women who comprise

[15] Sixteen pieces of land were consolidated in the proclamation of iSimangaliso Wetland Park in order to secure an open conservation area including from the Lubombo mountains in uMkhuze to the coastal plains of the Eastern Shores and Lake St Lucia to the sea that will support behavioural, nutritional, breeding and habitat requirements of animals (iSimangaliso Wetland Park Authority, 2008).

60 per cent of the workers employed. Ngcobo (2009) showed that the income earned through the programme was an important livelihood strategy. One of the respondents in Ngcobo's (2009, p. 57) research stated:

> 'Yes, the households around the park are getting help because they can now build houses and send kids to school. They also get help, especially the single women are the ones getting much help.'

Mngomezulu (2009) found that community-based contractors had experienced significant changes in their lives with the money earned through the land care work going to large assets such as houses and furniture, as well as business equipment such as vehicles. A woman respondent in Mngomezulu's (2009, p. 74) research said:

> 'As a mother, knowing that your children go to bed full at night is a life-changing experience on its own … life is very different now and it's a good. We used to buy secondhand clothes, books and anything that we could get from secondhand shops. But now we can afford like most people to buy clothes from Jet and Mr Price.'

The economic empowerment of women through this programme was also highlighted in Mngomezulu's research:

211

> 'I do feel like I am empowered and more in control, there are decisions that were so small that I couldn't make without my husband because he was the one making the money in the family but now I can decide and make decisions even without him, not to undermine him but to help him and the family. Building a house was one of them. I took the initiative and feel so empowered as well as sending my child to

700-year-old fish traps at Kosi Bay that are still maintained and used. iSimangaliso provides sustainable community access to natural resources. Many people rely on the park's mussels, fish and prawns.

Local people working to upgrade local roads: iSimangaliso's tourism and conservation infrastructure upgrades have created 22,650 jobs.

university. He always wanted to go but we couldn't afford this on his father's salary, the work I do at the park enabled me to make my child's dream come true.'

Between 2001 and 2008, to meet both biodiversity and tourism development goals, 1,503 mammals from seventeen different historically occurring species were reintroduced.[16] Through the park's Community-based Natural Resource Management Programmes, iSimangaliso provides sustainable access by communities to natural resources.[17] To support sustainable agriculture, the Authority established forty food gardens around the park in 2008, working with 900 gardeners, mainly women.

Sound ecological management and preservation of the park's outstanding universal value underpin the benefit flows, with tourism recognized as the region's lead economic driver, and empowerment and transformation goals cut across all aspects of iSimangaliso's commercial development work. World Heritage status and its new associated branding and marketing strategies have led to the steady growth of tourism, which has been boosted by improvements in infrastructure and access. iSimangaliso's efforts have had a very positive effect on tourism growth in the area.[18]

Land claimant groups are equity partners in three tourism accommodation facilities that provide for equity shareholding of 20–61 per cent. Tourism activity licences[19] are reserved for businesses where at least 70 per cent shareholding is community-owned and actively managed. Training programmes

[16] These include 40 elephant, 225 buffalo, 115 kudu, 60 white rhino and 8 cheetah.

[17] Many people rely on the park's mussels, fish and prawns; wood for building, fuel and carving; medicinal plants; grasslands for grazing; and *iNcema* (rush) for craftwork.

[18] A 2006 survey established that the number of tourist facilities had increased by 59 per cent around the park since the establishment of the iSimangaliso Wetland Park Authority. At the same time occupancy rates increased from below to well above the national average at the same time as a 20 per cent increase in bed numbers. A more recent survey (2010) indicates that despite the depressed global tourism market the region is holding its own and has not shown any decreases in tourism activity. Year-on-year visitor numbers to the park have increased (Zaloumis, personal communication, February 2012).

[19] These include scuba diving, boat cruises, horse riding, birding, turtle viewing, game drives and canoeing.

Aerial view of the vegetated coastal dunes and freshwater system of Lake Sibaya with the Indian Ocean in the background, showing the 'superlative natural phenomena and scenic beauty' of the park.

in tourism, hospitality and tour-guiding have included local people in the growth. These new partners benefit directly from conservation and in this way the park's outstanding heritage values have become tangible.

The Craft Programme provides focused support for successful income generation for local women through capacity-building, product development and selling crafts in a range of higher-value markets. Beginning in 2000, the Programme has supported twenty-four craft groups.

The Craft Programme is an important intervention for women, many of whom are not literate and will never gain access to formal employment in the tourism sector. By accessing higher-value markets, rather than roadside craft markets, incomes earned from craft production have increased substantially. Income earned through the Programme has enabled women to support their households:

'Living in grasslands area we had no idea that it could change our lives so much. After working with Product Developers we now know we can make a living using our natural materials as creative people. We are now able to pay our kids to school fees' (Mrs Mdlestshe, quoted in iSimangaliso Wetland Park Authority 2005, p. 2).

The park entry fees structure allows for equitable access, providing for free and discounted access, and there is a wide range of accommodation and facilities for day visitors. Schools and neighbouring communities are regularly invited. iSimangaliso's ongoing School Awards Programme attracted 373 submissions in its first two years.

The Rural Enterprise Programme provides sub-grants and technical assistance. A rural enterprise hub is being set up to provide enterprises with ongoing support. So far, ninety-eight enterprises have participated and twenty-eight have received grants.

iSimangaliso's support of the land reform process is a further mode of community benefits. Claimants become mandatory equity partners in tourism development and co-managers of the lands. In two conservation firsts, land claimant community representatives sit on the iSimangaliso Board, and the park's methods have informed national policy on the settlement of protected area land claims across South Africa.

213

In two conservation firsts, land claimant community representatives sit on the iSimangaliso Board, and the park's methods have informed national policy on the settlement of protected area land claims across South Africa.

Fifty artists from local communities have been trained in technical and business skills. Sculptors, painters, and dance and drama groups, through the Art and Drama Programmes, have received support and recognition for their openness to exploring the links between art and conservation. Samuel Mtshali, one of the iSimangaliso artists, states:

'I speak about the terrible events reported in our newspapers and television about rhino poaching. The animals are innocent. However, there are these "crocodiles" whose secret purpose is to end the existence of the rhino because they can get paid in euros or rands. I say it's a river without fish. It is sad to see blood in our parks. Those thousands of litres of blood that our generation sees daily will make them merciless in killing any living organism in future. The death of any animal is like losing a family member. Let's stop it for the next generation' (iSimangaliso Wetland Park Authority, 2011).

By 2012, iSimangaliso had supported forty-five young people from land claimant and neighbouring communities to study at tertiary institutions in the fields of conservation and tourism to develop skills for the future.[20] Students also participate in an annual work-based experience programme with the iSimangaliso Authority. According to Sinenhlanhla Mfekayi, 'working with iSimangaliso was helpful because as a tourism student, I get time to know my industry and interact with them. I also get the time to see the park as the first World Heritage site in South Africa. Now I feel and see the future in tourism' (iSimangaliso Wetland Park Authority, 2012).

iSimangaliso has made measurable progress toward its goal of becoming a 'People's Park' through development and rehabilitation. Thousands of park neighbours have benefited from employment via its land care and infrastructure contracts; rural enterprise programmes; sustainable agriculture and natural resource use policies; and its local economic development and training in tourism, leadership, environment education, art, craft and cultural heritage.

Lessons for other World Heritage sites

iSimangaliso's success in delivering benefits to local community groups and individuals is a result of its alignment with the South African government's macro-economic policy. By being part of the state's drive to deliver jobs and alleviate poverty, iSimangaliso has enjoyed strong political support for the World Heritage site that has enabled it to deal with external potential threats

---

[20] The Programme provides financial assistance through bursaries and loans, an academic support and development programme; an annual students' workshop; and in-service training assignments.

214

Local learners inspect a sea urchin at Sodwana Bay beach. The goal of the iSimangaliso Environmental Education Programme is that all students living in and around the park will visit iSimangaliso at least once in their school career.

such as mining on the periphery. Furthermore, this has enabled iSimangaliso to access funding for the conservation of the site and its rural development programmes. For example, expanded public works budgets have been secured to build infrastructure, create jobs and build the capacity of local contractors.

Through its 'Development to Conserve' approach, iSimangaliso has shown that local participation and community benefits are important aspects of building support for the conservation and protection of its World Heritage status, particularly in a context of poverty which leads to the unsustainable utilization of this natural resource, and a lack of economic opportunity in the region. The experience of being forcibly removed from iSimangaliso during the colonial and apartheid eras is within the living memory of the people. Landowners are now acknowledged by giving recognition to the cultural meaning of the landscape through the use of local names and interpretation.

Having established a track record of delivery, the iSimangaliso Authority is faced with the challenge of heightened expectation and the demand to expand and deepen its reach. Implementing a sustainable development strategy for iSimangaliso means that these high expectations will not be met. Managing expectations and building relationships of trust is an ongoing process that is not without its challenges and requires staff skilled at working with rural communities. The poor service delivery by local government and some other government departments in the region will remain a challenge for iSimangaliso. Large economic projects outside the park are required to deliver economic empowerment and jobs. Building partnerships with government agencies and departments, and NGOs seeking to alleviate poverty in the area, is therefore crucial.

As the benefits from iSimangaliso have increased, so have conflicts emerged in and between beneficiary groups. This has proved to be particularly challenging for the land claims leadership in the area and has required ongoing mediation support from iSimangaliso. Much of this type of conflict is concerned with controlling access to resources and gaining power through networks of patronage. Through participation and community benefits, attitudes towards conservation can change, showing that it is possible to 'develop to conserve'. It is important, however, to recognize that participation does not necessarily

Tourists watch hippopotamus at a watering hole in iSimangaliso Wetland Park.

lead to consensus; a wide range of interests exist and it is not possible to accommodate every point of view.

iSimangaliso has shown that recognition of World Heritage values is possible in circumstances where there are issues that may impact on the integrity of the site and that World Heritage status can help to secure the conservation of the site's outstanding universal value. Listing iSimangaliso as a World Heritage site has assisted with the process of resolving these issues by securing political and financial support, and raising the status and profile of the park.

216

Attitudes towards conservation can change, showing that it is possible to 'develop to conserve'. It is important, however, to recognize that participation does not necessarily lead to consensus; a wide range of interests exist and it is not possible to accommodate every point of view.

Benefits beyond boundaries

iSimangaliso Wetland Park has many achievements showing that the integration of the conservation of outstanding natural heritage values and socio-economic development and redistribution is possible. Perhaps most importantly, the understanding of what protected area management can mean has been transformed. Many local communities were removed from their land in the past for the creation of what is now the park, and negative perceptions of conservation were created. The park's commitment to balance the needs of conservation, social justice and public interest in a sustainable way has contributed to the restoration of local support for protected area management. Its successes in spreading 'benefits beyond boundaries' has shown that outstanding universal value is not the preserve of the few and all can benefit from conservation of heritage.

The park has fulfilled its responsibilities to the UNESCO World Heritage Committee against the stringent ecological guidelines to which it is obligated, and been cited by UNESCO as a leading example of the 'benefits beyond boundaries' conservation model. On the occasion of the park's tenth anniversary, the Minister of Water and Environmental Affairs congratulated iSimangaliso on turning the park into 'a driver of economic upliftment and social regeneration' – she described it as an approach which had 'effectively changed the face of conservation'.

18

Participatory methodologies and indigenous communities – project-based learning: Sian Ka'an, Mexico

JULIO MOURE AND JESSICA BROWN[1]

Where the Sky is Born[2]

Inscribed on the World Heritage List in 1987, Sian Ka'an was first recognized as a Biosphere Reserve in 1986. It is the largest protected area in the Mexican Caribbean, encompassing terrestrial and marine environments of high biological diversity with unique geological features. Its location on a partially emerged coastal limestone plain has resulted in unique geological features, such as sinkholes (*cenotes)* and underground rivers, important for their high biodiversity and species endemism. Its 651,029-ha area[3] encompasses a diversity of coastal and marine environments representative of the Caribbean Sea and the Yucatán Peninsula, including sandy beaches, rocky beaches, sand dunes, mangroves, shallow bays and coral reefs.

Sian Ka'an protects a 110-km portion of the Meso-American Barrier Reef, the second largest in the world, rich in marine biodiversity, including 52 species of reef fishes. On the terrestrial side, as part of the Sian Ka'an-Calakumal connector, it contributes to connectivity across the forested landscape within the Meso-American Biological Corridor. In addition to high floristic diversity and the presence of many endangered mammal species, the Biosphere Reserve supports the second largest community of aquatic birds in Mexico and is a key part of the migratory bird corridor between North and South America (López-

---

[1] Jessica Brown, a global consultant with the COMPACT initiative; Julio Moure, Regional Coordinator, COMPACT-Mexico.

[2] *Sian Ka'an* may be translated from the Yucatec Maya language as 'Where the Sky is Born' or 'Gift from the Sky'.

[3] http://natoural.conanp.gob.mx/SNIAT_en.html

Ornat, 1990). There are 346 bird species registered in the reserve, including resident and migratory species (MacKinnon, 1992). With more than 300,000 ha of aquatic environment it supports the largest crocodile habitat found in any of Mexico's protected areas (Lazcano-Barrero, 1990) and is particularly rich in amphibians and reptiles. A preliminary listing of over 100 mammal species found in the reserve includes manatees, dolphins, four species of whale and thirty-nine species of bat (Instituto Nacional de Ecología, 1993).

The region, known as 'the heart of the Mayan culture', is rich in the cultural heritage of its past and present-day inhabitants, in particular indigenous peoples. The high degree of biodiversity found conserved is partly a legacy of the traditional knowledge and practices of the Maya people and their management of the landscape over the centuries. Sian Ka'an is located in the ancient Mayan regions of Cohuah and Uaymil, probably inhabited during the pre-Classic and Classic periods. There are twenty-three known archaeological sites of pre-Hispanic culture in the Biosphere Reserve, and discoveries of human remains, ceramic pieces, and other artefacts have been dated up to 2,300 years old. Today, small communities in and around the reserve are predominantly of Mayan origin and a number of indigenous languages are spoken in the area. The population is estimated at 2,000, with most settlements concentrated in the coastal regions. The Mayan communities hold possession of the land in the form of *ejidal* land tenure.

> The high degree of biodiversity found conserved is partly a legacy of the traditional knowledge and practices of the Maya people and their management of the landscape over the centuries.[4]

This case study presents the work of the Community Management of Protected Areas Conservation Programme (COMPACT) in Sian Ka'an.[5] It explores the use of a highly participatory methodology based on dialogue that has forged partnerships with local communities to improve conservation of the World Heritage site, while improving local livelihoods and helping to stem the loss of Mayan languages and culture. Key project elements include sustainable fisheries management, community-based tourism activities and carbon capture. Experience with reviving traditional knowledge in farming, handicrafts and apiculture is discussed, along with empowering community-based networks to market local products under a common brand.

Context for sustainable development

Although it is in the least-developed part of Quintano Roo, the Sian Ka'an Biosphere Reserve and World Heritage site still faces a number of threats.

[4] Text inspired by the publication of Boege 2008.
[5] An initiative of the UNDP/GEF Small Grants Programme (SGP) and United Nations Foundation (UNF), COMPACT-Mexico has been working in close partnership with communities in the Sian Ka'an landscape and seascape since 2000.

218

The Sian Ka'an region, known as 'the heart of the Mayan culture', is rich in the cultural heritage of its past and present-day inhabitants, in particular indigenous peoples.

Unregulated tourism development, overfishing, forest fires, cultivation of coconut in the coastal dunes, and the uncontrolled extraction of resources are some of the main activities threatening the site. Cancún has been transformed from a fishing village to the largest tourism destination in Mexico. Ongoing development along the coast contributes to the contamination of the water and is altering the hydrology of the area, compromising the integrity of the estuarine, mangrove and coral reef communities.

New developments in agriculture pose further threats to the region's landscape. Growing reliance on intensive industrial inputs, 'improved' seeds (i.e. hybrid and transgenic), fertilizers and pesticides, and the use of machinery are having a major impact on land use, contributing to soil erosion, groundwater contamination, and the loss of biodiversity and agro-biodiversity. At global level the loss of ecosystem services formerly provided by these natural systems is significant (Boege, 2002).

The present-day Mayan culture, with all the contradictions and challenges it faces, possesses a rich heritage of knowledge and management practices (traditional ecological knowledge or TEK). Whereas the concept of biodiversity is very recent, the practices of its sustainable use and conservation by indigenous peoples, such as the Maya, span millennia. The territories these peoples inhabited contain enormous biodiversity that has contributed to the global inventory. In short, the living cultural heritage of this landscape and its inhabitants is an inextricable part of its global significance. But external pressures are threatening local culture and livelihoods, in particular of the indigenous peoples living in the Sian Ka'an landscape. Financial, ecological and social debt from the market economy is resulting in extreme poverty, the loss of indigenous language and culture, out-migration from the region and unemployment.

The Mayan cultures have lived for years with the ecosystems of the Sian Ka'an landscape, and have co-evolved with them, choosing to use some plants and animals, cultivating others, so that their practices have transformed the landscape and its biodiversity (Toledo and Barrera-Bassols, 2008). With the selection of wild species came the development of cultivated plants that were distributed worldwide and are now the basis of the global food system. Indigenous production systems have long sought to optimize their use of local resources and adapt to environmental conditions, based on shared knowledge, technologies and ways of organizing work according to the preferences and values of the group (Bonfil Batalla, 1994). Importantly, this experience is not only restricted to food. Living alongside the biodiversity of the region has required these communities to develop complex ways of using the plants, insects and animals around them for food, medicine, clothing and shelter. Bringing together scientific and indigenous knowledge systems is an important challenge for COMPACT, ensuring that the initiative respects the natural ecosystem while helping to meet basic human needs.

Participatory methodology and governance

The COMPACT programme was launched in Sian Ka'an in 2000, building on the substantial experience of the UNEP/GEF Small Grants Programme's (SGP) prior decade of work in the Yucatán Peninsula. The methodology relies on participatory planning through three closely linked elements: a *baseline assessment*, which serves as the foundation for a *conceptual model* and *site strategy* that guide COMPACT's work at site level (see Brown et al, 2010). The COMPACT coordinator over the first seven months conducted numerous meetings with community-based groups, NGOs, environmental authorities, local authorities and academics to identify challenges and help to frame how communities could address them.

Participants in this consultative process identified as a central challenge: to provide livelihood opportunities for local residents while resisting the negative effects of the very rapid rise of tourism along the coastline; and to develop sustainable ecotourism approaches to benefit local communities as an alternative to 'selling out' areas of coastline to large-scale private developers. As a result of this collaborative process, a bilingual document was produced by COMPACT Sian Ka'an in Spanish and Mayan, using simple language and drawings by a local artist. It served as a starting point to explain and understand the goals and operations of the programme.

This approach is founded on principles of empowerment and endogenous development, such as those articulated by Paulo Freire (2005). The programme seeks to create answers to problems in dialogue with people in order to find, in their plain language, the seeds of solutions to multi-faceted problems that emerge from a long history of marginalization. In this view, 'knowledge is not

Medium-altitude semi-evergreen forest represents the climax vegetation (where growth is undisturbed) in non-flooded areas. Some 120 trees and shrubs have been recorded, including larger trees.

transmitted, rather it is "under construction", meaning the act of education is not a transfer of knowledge, but rather the enjoyment of building a common world (Souza, 2011).

Each step is defined through a diagnostic and collective planning process that creates a framework for responsibility and cooperation among grass-roots groups, participating NGOs and other actors. The aim is to trigger new attitudes, raise awareness and strengthen self-development. Capacity-building is seen as a process of lifelong learning – from practice to knowledge, from knowledge to vision, and from vision to action. Collective learning encourages teamwork and transforms competition into emulation, alongside the fundamentals of creativity, respect and commitment.

The local coordinator is responsible for planning and implementing the COMPACT programme, and serves as a key link between communities, diverse stakeholders, and the SGP Country Programme and its National Steering Committee. Advisory structure at the local level parallels that of SGP, operating in a decentralized, democratic and transparent manner. A local selection committee of ten people with expertise in different areas of the programme makes decisions on funding of projects in coordination with the Local Coordinator and the SGP National Steering Committee. The committee members are volunteers working in thematic clusters such as forestry, fisheries, tourism, Mayan culture and bee-keeping, to advise the Local Coordinator on programme planning in these areas and to offer their expertise to COMPACT grantees.

Key areas of work in Sian Ka'an

Over the past decade COMPACT has financed over ninety small grants supporting projects in and around the Sian Ka'an Biosphere Reserve and World Heritage site in three thematic areas: the coast, the forest, and the preservation of Mayan culture. A fourth line of work, environmental education and technical support, serves as the 'fishing rod', supporting the development of skills in intercultural dialogue.

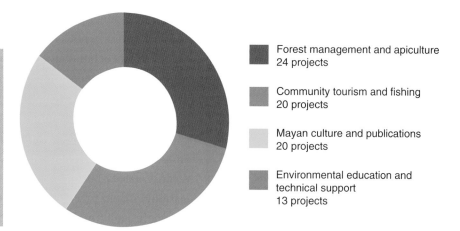

Thematic areas
receiving grants from
COMPACT

Forest management and apiculture
24 projects

Community tourism and fishing
20 projects

Mayan culture and publications
20 projects

Environmental education and
technical support
13 projects

Coast

Based on the successful experience of the Vigia Chico fishermen's cooperative in Punta Allen, sustainable practices of lobster fishing (e.g. the use of
shades instead of traps and protection of nursery areas) have been extended
to numerous other fishing cooperatives in communities such as Maria Elena
and Azcorra in nearby Punta Herrero. The experience of the Integrated
Association of Lobster Fishermen, CHAKAY, has been extended to three more
cooperatives (Banco Chinchorro), linking two biosphere reserves. To help
these groups with joint marketing of lobsters, COMPACT has worked with
WH-LEEP (World Heritage Local Ecological Entrepreneurship Programme)

General view of the
coastal wetland area.
Located on the east coast
of the Yucatán Peninsula,
the site is bounded by the
Caribbean Sea and the
barrier reef.

Locals net-fishing who compete with pelicans for fish. The site is rich in marine biodiversity, including 52 species of reef fishes.

223

to provide small grants to support selection and packaging of the lobsters, including developing an origin label. In its first year the cooperatives reached their target of jointly marketing about 10 per cent of their production.

In an important marine conservation initiative extending beyond the boundaries of the reserve, COMPACT has worked over the past decade with three fishing cooperatives on protection of fish aggregation and spawning zones within the Sian Ka'an Biosphere Reserve. In the second stage of the project these partners are analysing the potential to create marine protected areas, likely to take the form of fishing refuges, or 'no-take' zones (lasting at least five years), a proposal that is supported by key local fishing cooperatives.

Another key area of work is coastal tourism. Punta Allen is the principal point of attraction for tourism along the coast of Sian Ka'an, attracting between 80,000 and 100,000 visitors annually. Through ten years of project support, four tourism organizations in the community formed the Punta Allen Alliance, in collaboration with protected area authorities and international environmental organizations committed to conservation programmes that benefit both people and nature, such as Rare. This partnership has been crucial in avoiding conflicts and maintaining consistent prices for tourism services and products offered in the community.[6] An important result is ensuring

[6] http://www.puntaallenalianza.com/

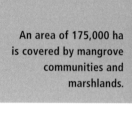

An area of 175,000 ha is covered by mangrove communities and marshlands.

that the majority of the benefits remain with the communities. In cooperation with the Punta Allen Alliance, support has been made available for a new tour guide training course for women, and a cooperative (Orchids of Sian Ka'an) offering new initiatives for visitors interested in cross-cultural exchange and nature-based tourism.

Forest

Promoting apiculture has proven to be one effective way to help maintain forest cover while improving the quality of life for people in the region. Several organizations are making the transition from conventional collecting to organic honey bee-keeping and have successfully obtained organic certification, such as Flor de Tajonal, a certified cooperative that sells between 150 and 200 tons of honey annually and is leading a process of landscape-level cooperation among various communities in the Mayan region. Plans are under way to establish an apiculture school where young people from local communities can study the theory and practice of organic bee-keeping. At the same time a reforestation project involving native honey plants is helping to boost productivity.

Sustaining Mayan culture is an important part of the COMPACT interventions in the Sian Ka'an landscape.

In 2007, the carbon capture project was initiated in the Ejido de Felipe Carrillo Puerto focusing on a 1,230 ha area. The pilot project, Much' Kanan K'aax,[7] has grown and become a centre for learning and sharing of experiences on this subject for the entire Yucatán Peninsula. In the next stage of the project, COMPACT will finance the certification of bonds (through the *Plan Vivo* Foundation) and will support training and capacity-building of local communities on the topic of REDD+.[8] A long-term forest partnership is in development, involving five *ejidos* working in a 200,000 ha forest area to improve stewardship and secure timber certification.

Since 2008, with the support of the United Nations Foundation and COMPACT, a partnership involving two NGOs and representatives of eight community groups is jointly marketing handicrafts, including items made from wood, seeds and rattan (Non-timber Forest Products), as well as embroidery and hammocks, under a common indigenous trademark. All the participating groups come from Mayan communities in the area and already have a tradition in the development of handicrafts. At present 139 artisans from fifteen communities are collectively marketing their handicrafts under the *Ak Kuxtal* label.[9]

225

Mayan culture

An integrating element across all biophysical interventions in the Sian Ka'an landscape is to sustain the Mayan culture. The Maya Intercultural University of Quintana Roo is a key partner. Some 600 young people from local communities now study there, pursuing careers in fields such as agro-ecology, community health, Mayan language and culture, alternative tourism and

---

[7] http://muchkanankaax.com/

[8] Reducing Emissions from Deforestation and Forest Degradation (REDD) is a UN collaborative programme to create a financial value for the carbon stored in forests, offering incentives for developing countries to reduce emissions and invest in low-carbon paths to sustainable development. REDD+ goes beyond deforestation and forest degradation, and includes the role of conservation, sustainable management of forests and enhancement of forest carbon stocks.

[9] Ak Kuxtal means 'Our Life' in Maya, see http://www.kuxtalsiankaan.com/ak-kuxtal.php.

municipal management. Elements of work in this area include publications in the Mayan and Spanish languages, presenting biological information as well as symbolic representations, stories and legends. The COMPACT programme has funded nine bilingual publications now found in 510 community centres and schools in the region.

Strengthening of local organizations includes those concerned with traditional medicine, language and culture. Recovery of native seed stock with twenty communities addresses the urgent need to conserve seeds and plants adapted to growing in the region, especially those most important for human nutrition. Research and training in techniques of using natural dyes is based on the experience of several states with a strong indigenous presence (Oaxaca, Chiapas and Quintana Roo), and in collaboration with people from local Mayan communities, a manual was published showing how to produce nine colours with natural plants. The next stage will be the production of fabrics, hammocks and other products using natural dyes.

Impacts on indigenous and local communities

Between 2000 and 2005, poverty rates fell dramatically in Punta Allen, Punta Herrero and Maria Elena, all communities within the reserve. The percentage of households experiencing nutritional poverty declined from 32.16% to 5.38%. The rates of poverty of capabilities (skills) and poverty of wealth also fell from 50.29% to 8.6% and from 85.38% to 22.58%, respectively … [these poverty indicators related to] nutrition, skills and capital are lower in these locations as compared to state and national averages (UNESCO Mexico, 2009).

Activities relating to fisheries, ecotourism and bee-keeping have resulted in significant increases in household income in communities, distributed effectively through families (Table 1). Income-generating activities, linked to certification of good ecological practice, have resulted in an increase of income in those households reached by the projects, estimated in the range of US$1,000,000 in an average year.

Table 1 Relative increases in income in projects financed by COMPACT

Project	Increase in income	Source of income
Lobster fishery	30%	Sales of lobster
Apiculture	20%	Sales of organic honey, mainly to Europe
Forest management	20%	Sales of certified wood
Community tourism	20%	Reduced consumption of gasoline using fuel-efficient motors
Handicrafts production and commercialization	20%	Sales of community products made from sustainably managed resources under a common brand and label of origin
Organic agriculture	10%	Sales of a portion of organic crops, with remainder for family consumption

Promoting apiculture has proven to be one effective way to help maintain forest cover while improving the local quality of life.

Productive activities such as fisheries, forest management and apiculture have traditionally been the domain of men, and they remain the main participants in these activities, with women participating in only about one-third of the COMPACT projects in these areas.

Between 2000 and 2005, poverty rates fell dramatically in Punta Allen, Punta Herrero and Maria Elena, all communities within the reserve. The percentage of households experiencing nutritional poverty declined from 32.16% to 5.38%. The rates of poverty of capabilities (skills) and poverty of wealth also fell from 50.29% to 8.6% and from 85.38% to 22.58%, respectively… [these poverty indicators related to] nutrition, skills and capital are lower in these locations as compared to state and national averages (UNESCO Mexico, 2009).

The benefits to ecosystems and the landscape as a whole are shown in Table 2.

What began as small projects linked to programme priorities are now becoming progressively organized as clusters. At present, different stakeholders and partners in the region are working together to develop plans along the following lines of work: Mayan culture, fisheries, tourism, forestry management and apiculture. In the third phase partnerships are being forged, based not only on planning in common, but also making cooperative agreements and organizational commitments. These include: partnerships in fishery (e.g. Chakay, fishing cooperatives in Sian Ka'an and Banco Chinchorro), partnerships in community-based tourism (e.g. Punta Allen Alliance), and partnerships for forest protection (alliance of forest *ejidos*). These alliances, started at local level, are now extending their reach across the state and expanding to focus on the entire Yucatán Peninsula. Examples include the Alianza Kanan Kay, which is concerned with the entire coast of Quintana Roo, and Alianza Itzinkab, which is concerned with the forest of the peninsula.

227

Project-based learning

Among the lessons learned from a decade of work in the landscape of Sian Ka'an are the following. At each stage it is crucial to encourage dialogue

Sian Ka'an lies on a partially emerged coastal limestone plain which forms part of the extensive barrier reef system along the eastern coast of Central America.

Table 2 COMPACT beneficiaries and areas under conservation

Eighty-six projects financed by COMPACT*	Total amount US$1 952 530	Origin of funds 75% GEF 25% UNF	Average per project US$22 704
Beneficiaries (including environmental and cultural education)	Women: 5 962	Men: 7 427	Men: 55.5% Women: 45.5%
Beneficiaries in productive projects	Women: 1 461	Men: 4 501	Men: 66.7% Women: 33.7%
Hectares under management (approximation)	Marine: 120 000 ha (two bays of Sian Ka'an, fisheries and community tourism)	Forest and land: 130 615 ha (forest management, apiculture and organic agriculture)	

*An additional five projects were financed by WH-LEEP during 2011.

Projects in the communities mature at their own speed and gain traction when people see the results: that these methods of conservation and collaboration have allowed them to improve their incomes and maintain their resources over the long term.

directly among participants in order to share the positive, analyse the difficulties and challenges and then adjust what is not going well. This reliance on dialogue supports an adaptive management approach to developing projects. It is important to create partnerships and alliances that combine the efforts and benefits of related sectors.

Based on the successful experience of the Vigia Chico fishermen's cooperative in Punta Allen, sustainable practices of lobster fishing have been extended to numerous other cooperatives.

Conclusions

In its 12 years of work in the Sian Ka'an World Heritage Site, COMPACT has fostered a landscape-level laboratory for initiatives that advance sustainable development, sustain indigenous culture, and build social capital. It has demonstrated tangible progress in improving livelihoods and enhancing conservation, in areas ranging across fisheries, apiculture, handicrafts, community-based tourism and forestry. Using a participatory and community-driven approach, COMPACT has been able to open up new perspectives and attitudes among local communities and other stakeholders. This work relies on partnerships with a broad range of stakeholders and would not be possible without the cooperation of partners in government, academia, business and the NGO sector.

As it develops and expands its partnerships, COMPACT has the potential to continue to grow and amplify its impact in the Sian Ka'an landscape and the wider region of the Yucatán Peninsula in the coming years. At national level the Mexican Government is drawing on the experience of the UNDP/GEF Small Grants Programme in the Yucatán Peninsula (including COMPACT's work in Sian Ka'an) to develop landscape-level programmes in protected areas in other parts of Mexico. Thus COMPACT's impact is reaching beyond the Sian Ka'an World Heritage site to other regions of Mexico. By piloting this integrated approach over an extended period in a globally significant landscape, COMPACT is serving as a model with national and international relevance.

19 Village on the winding river: Historic Villages of Korea: Hahoe and Yangdong, Republic of Korea

AMARESWAR GALLA[1]

Clan villages

Historic villages in the Republic of Korea are of diverse forms, but 80 per cent of them are clan villages. The genesis of this type of settlement dates from the late Goryeo dynasty (918–1392) and they became typical of Korean villages from the latter part of the Joseon dynasty (1392–1910). Clan villages are now on the decline with the rapid urbanization and industrialization of Korea. Hahoe and Yangdong, the clan villages inscribed on the World Heritage List in 2010, are traditional and fully express the academic and cultural achievements of the Joseon dynasty. The agencies responsible for the properties are the Cultural Heritage Administration at central government level, Gyeongsangbuk-do at provincial level and Andong and Gyeongju cities at local government levels.

These are two of the best preserved and representative examples of clan villages, and in their siting, planning and building traditions they make an exceptional testimony to the Confucianism of the Joseon dynasty, which produced settlements that followed strict Confucian ideals over a period of some 500 years. The two villages faithfully adhere to the *pungsu* principle (traditional siting principle, *feng shui* in Chinese), in village construction. One sits along a river and the other at the foot of mountains, thus demonstrating best examples of desirable clan village locations. They are among the very few examples of

[1] Keynote speaker and resource expert in the Hahoe workshop on Integrated Heritage Management organized by the Mayor of Andong, Cultural Heritage Administration and ICHCAP of the Korean Heritage Foundation. February 2011. The author also drew on both published and unpublished material from Hee Ung Park and Hyosang Jo, Cultural Heritage Administration, and Professor Okpyo Moon, Academy of Korean Studies.

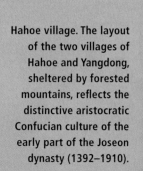

Hahoe village. The layout of the two villages of Hahoe and Yangdong, sheltered by forested mountains, reflects the distinctive aristocratic Confucian culture of the early part of the Joseon dynasty (1392–1910).

231

intact traditional Korean clan villages that have kept their original spatial layouts featuring three areas for productive, residential and spiritual life, which exceptionally are still functioning.

The village complexes of Hahoe and Yangdong reflect the impact of the Joseon dynasty that profoundly influenced the development of the Korean peninsula over five centuries. The villages and particularly the neighbourhoods of *yangban* (literati or nobles) and commoners' houses, and their overall and individual planning, reflect the precepts of this dynasty in terms of its social structures and cultural, literary and philosophical traditions, as well as its power and influence.

The two villages feature a number of historic buildings, including houses, shrines, study halls, Confucian academies and pavilions that are among the oldest and most remarkable in style to be found in the traditional villages of the country. They have preserved many ancient documents and works of art from the prominent academic and cultural activities of Joseon period Confucian scholars and traditional family rituals and unique communal events still take place there.

This case study focuses on Hahoe village as it has become a project for Korea in addressing the integrated management of diverse heritage resources, tangible and intangible, that are part and parcel of the World Heritage site. Moreover, Yangdong is currently the focus of development, while Hahoe

Byeongsanseowon Confucian Academy of Hahoe. As the custodians of their heritage, Hahoe villagers have had to struggle to resist the external appropriation of their culture and to reclaim the cultural stewardship of their village.

has received considerable attention since 1984. Hahoe also has a significant number of community members living in the World Heritage site. It is a well-known village for *yangban*. The residents are mainly from the *Ryu* lineage, the descendants of *Ryu Seong-ryong*. The Nakdonggang River meanders around the village in an S-shaped curve. It has retained the natural elements of the 'floating lotus shape' for six centuries, compatible with *pungsu*.

Hahoe was designated by the Korean Government as an Important Folk Property in 1984 due to the large number of traditional upper-class houses built there during the Joseon dynasty. The Hahoe Mask Dance, traditionally performed by the servant classes, was made a state designated Intangible Cultural Asset in 1980. Even though the village had received considerable recognition, it was in decline as the local population was ageing fast while younger residents were emigrating in search of a better quality of life. However, the aspirations for World Heritage status for the village and the subsequent developments after inscription have proved to be turning points, breathing new life into Hahoe.

Continuity of the village

The meaning of sustainable heritage development warrants an understanding of the layers of significance embedded in the recent history and heritage of

Hahoe *Byeolsingut Tallori*. The Hahoe Mask Dance, traditionally performed by the servant classes, designated an Intangible Cultural Asset by the state in 1980. The performance has become a major heritage tourist attraction at the World Heritage site.

Hahoe village. The beginning of nationwide recognition was in the 1970s for its significance as a centre of Confucian tradition. One of the aspects of living heritage is that it can be used by a range of actors in societal development. The Hahoe *Byeolsingut Tallori* Mask Dance Drama was revitalized as a symbol of anti-establishment movements in Korea in the 1980s. It profiled the village nationally and attracted many dissidents and students to Hahoe. This new-found heritage consciousness made Hahoe a tourist destination from the early 1990s, resulting in the commodification of local heritage and threatening its integrity and authenticity.

> As the custodians of local heritage, Hahoe villagers have had to struggle to resist the external appropriation of their culture and to reclaim the cultural stewardship of their village.

As the custodians of local heritage, Hahoe villagers have had to struggle to resist the external appropriation of their culture and to reclaim the cultural stewardship of their village.[2] In the tension arising between the centripetal forces of localization and self-awareness and the centrifugal forces of globalization and external impacts, World Heritage status and appreciation of Hahoe's outstanding universal value have become mediators to minimize the negative impacts and maximize the opportunities offered by responsible tourism as a vehicle to sustainable development.

In this context, while the nomination to the World Heritage List was led by external disciplinary specializations and concerns, the inscription has had a profound effect, not only on Hahoe but on Korean heritage management. The

[2] O. Moon, 2005, Hahoe: the appropriation and marketing of local cultural heritage in Korea, *Asian Anthropology*, Vol. 4, pp. 1–29.

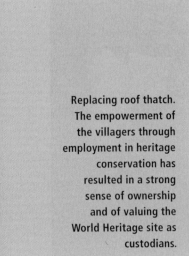

Replacing roof thatch. The empowerment of the villagers through employment in heritage conservation has resulted in a strong sense of ownership and of valuing the World Heritage site as custodians.

Recognition of the skills and knowledge of the local communities has come to guide all conservation measures, and respect for them is crucial to all cultural experiences provided.

current engagement to safeguard the outstanding universal value is informed by realization on the part of the national, provincial and local authorities that an integrated management approach to both tangible and intangible heritage is imperative to viability and sustainability. Recognition of the skills and knowledge of the local communities has come to guide all conservation measures, and respect for them is crucial to all cultural experiences provided.

The empowerment of the primary stakeholder communities, the villagers in their diversity of class and occupational contexts, has resulted in a strong sense of ownership and of valuing the World Heritage site as custodians. The opportunities for jobs and economic development provided through the growth in local GDP have softened the traditional hierarchy of political status to one based on actual and increasingly equitable status. The effect has been to stem the outward migration of local young people to the cities and contribute to the well-being of the older generation that was being left behind. Census data show that the number, 166, of registered households in 1964 was reduced to around 100 with a 75 per cent fall in the local population by 1999, when the village received the largest number of visitors ever following publicity from the visit of Queen Elizabeth II.

Intergenerational transmission and responsibility for heritage conservation have become important among all classes in the village. Hahoe is now a symbolic project for Korea, where the anticipated rural–urban migration will result in more than 80 per cent of the population living in about twenty urban conglomerations. Valuing rural life as the mainstay of traditional Korean life

Pickled vegetables (kimchi) prepared for the winter in Hahoe. Valuing rural life as the mainstay of traditional Korean life has become a concern at all levels of government.

235

has become a concern at all levels of government. Moreover, villages such as Hahoe have become educational resources for schools in order to teach Korean history and culture.

Hahoe governance

The stabilization of the local population and the promotion of participatory democracy called for the creation of new structures for governance of the village. In the inscription process, the World Heritage Committee requested that the State Party address the issue of implementation of a coordinated management system for the two component sites. In response, a Historic Village Conservation Council was launched in January 2010 with its legal grounding in Gyeongsangbuk-do's Ordinance on Conservation, Management and Support for World Heritage, to take charge of coordinated management for the serial property, and is now in full operation both legally and administratively. The Council holds regular quarterly meetings and extraordinary meetings as deemed necessary, in accordance with its Operational Regulations. Gyeongsangbuk-do province serves as its secretariat, organizing meetings and implementing decisions. As a semi-public organization with representatives from central, provincial and civic governments working with village residents

Fire-fighting exercise in Hahoe village. Recognition of the skills and knowledge for emergency preparedness by the local communities has come to guide all conservation measures, especially in the World Heritage site, where most of the houses are wooden or have thatched roofs.

and experts, the Council ensures continuing and systematic participation of the residents as an integral part of the heritage. At the operational grass-roots level, each village functions as an autonomous entity, sharing knowledge and experiences with each other and other clan villages in Korea.

In the lead-up to the inscription of the clan villages, the focus was on tangible heritage and inputs were mainly from external agencies. However, the conservation of the site's outstanding universal value, authenticity and integrity posed new challenges, gradually leading to cooperation and coordination at municipal, provincial and national levels. Hahoe was the first of the two villages to gain support from all the agencies involved in securing a sustainable future for its declining resident population.

Taking ownership and benefits

Hahoe village and its development as a national symbol of Korean traditional culture, demonstration project for educational resources and destination for heritage tourism has contributed to local community benefits in several ways. A current benefit analysis exercise is being undertaken and the following is an indication of the trends.

Local intangible knowledge, illustrated through the tangible heritage of structural and environmental resources, is the main tourist attraction. Visitor experiences are facilitated through the creation of products informed by the stories, performances and skills of the carriers and transmitters of living

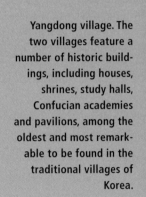

Yangdong village. The two villages feature a number of historic buildings, including houses, shrines, study halls, Confucian academies and pavilions, among the oldest and most remarkable to be found in the traditional villages of Korea.

heritage. Visitor numbers have increased substantially from the base line of 374,391 in 1998 to over a million within a year, stimulated by the royal visit and then averaging about 900,000 per year until 2011. Domestic visitors still comprise over 97 per cent of the total, although international visitors have increased from 5,488 in 1998 to about 25,000 in 2011.

The past decade has seen considerable government investment in Hahoe. Increased visitation and government support for conservation and infrastructure development have also led to increased prosperity for local households. This new-found affluence and job opportunities have become an incentive for young people to stay in the village or even move back from Andong and Seoul following their education and relevant training. Hahoe, along with Yangdong, is among the six 'Folk Villages' designated by the Cultural Heritage Administration to be protected by law. This ensures financial support from central and local government budgets, which constitutes Hahoe's main source of revenue for conservation and maintenance.

Expenditure supported by the government is limited to the physical conservation of the village, especially the repair and restoration of houses to maintain their original form, expansion of infrastructure, basic facilities for tourism, and the design of comprehensive plans for overall improvement of the village. In the period between 1984 and 2008, Hahoe received US$20.1 million as an initial investment from central government. In the past decade government funding has amounted to an average of US$1 million per year.

Aerial view of Hahoe village. Hahoe was designated by the Korean Government as an Important Folk Property in 1984 due to the large number of traditional upper-class houses built there during the Joseon dynasty. This laid the foundations for inscription on the World Heritage List and subsequent measures to conserve the site.

Gyeongsangbuk-do Provincial Developmental Plan is focused on Hahoe village as an integral part of the Confucian cultural region. It has invested US$40.1 million in the construction of an entrance to the village and tourist infrastructure including parking lots and a traditional shopping mall. The province also invested US$5.4 million in the construction of the Hahoe Village Route to improve the circulation of visitors and minimize negative impacts on residents' quality of life. This route connects the village with the local Byeongsanseowon Confucian Academy and Bongjeongsa Buddhist Temple.

The Culture and Arts Promotion Fund and the Lottery Funds also provide for employees in the tourism programmes in and around Hahoe. Some of these programmes are integrated as part of the wider festivals, educational and tourism-related activities benefiting from the substantially larger population and financial strengths of Andong City. This was kick-started in 2004 with a US$1 million subsidy to promote traditional boat trips and a river firework event and about US$3 million for the Hahoe Mask Dance Drama. Ongoing subsidized programming includes Confucian cultural experiences, ancestral rites and ceremonies to welcome guests, as well as the Mask Dance.

The local government provides a subsidy of over 40 per cent of the income generated from admission fees, which averages about US$3 million per year. These funds are a direct investment in the World Heritage property as the subsidy is given to and controlled by the Village Conservation Association. In addition, a similar proportion of revenue comes from parking charges and rental fees from the new store complex at the entrance to the village, the management of which Andong city has entrusted to the Conservation Association.

In addition to financial support for the historic villages, the central government provides a separate subsidy which is directly paid to families living in and managing the ancient houses. Each of these households receives about US$300–400 a month. About a dozen families in Hahoe Village benefit.

Continuing learning

The conservation of Hahoe village offers a valuable insight into local community engagement, especially the value of traditional village officials. Outside expertise relies on specific heritage elements, tangible or intangible, but village community culture requires a neighbourhood approach that brings all elements together.

Several lessons from the current experiences of Hahoe village are worth sharing and provide a focus for comparative studies in World Heritage conservation. Optimal benefits to the stakeholder communities can facilitate the continuity of the outstanding universal value through local ownership.

The first lesson is the inclusion of village or local community culture in conservation. The conservation of Hahoe village offers a valuable insight into local community engagement, especially the value of traditional village officials. Outside expertise relies on specific heritage elements, tangible or intangible, but village community culture requires a neighbourhood approach that brings all elements together.

The promotion of participatory democracy by bringing together the local communities in and around the World Heritage site and stimulating community-based development through heritage tourism contributes to poverty alleviation without compromising the integrity of heritage resources. It is important to conduct community and wider stakeholder benefit analysis so that the implementation of programmes and projects ensures economic and social benefits to the primary stakeholders.

World Heritage as a catalyst for sustainable development promotes responsible tourism based on heritage resources that are non-renewable. This requires the development of appropriate methodologies for systematic cultural mapping of heritage resources: tangible and intangible, movable and immovable, natural and cultural, creativity, and collaborative partnerships between the local communities and the best possible expertise for safeguarding the outstanding universal value of the site. The government's initial investment has facilitated this process.

The conservation and management of Hahoe village has been made possible only through the active participation of the residents. Top-down funding could easily lead to an external perspective in conservation, which not only marginalizes the heritage values of the residents themselves but leads to a lack of awareness among residents on dealing with conservation in the face of rapid change. The primary stakeholder community plays an important

239

role in preventing or delaying the changes undermining heritage values. They are the key beneficiaries of external investment and growth from tourism. Environmental change and development according to the organic characteristics of the village are now considered.

The way forward

Hahoe's experiences of the past decade are informing future developments. The Second Comprehensive Conservation and Management Plan will reflect learning outcomes from the primary plan. The focus is on the whole of the serial property including Yangdong village. A study is being conducted on the two villages that will include detailed guidelines on repair and restoration, landscape protection, fire prevention and implementing the recommendation made by the World Heritage Committee. A separate study on Tourist Impact Assessment is researching current trends in visitors and formulating tourism impact monitoring indicators while assessing visitor capacity of the villages. An integrated website for the Historic Villages as a World Heritage property is also under way, offering up-to-date information on the heritage in various languages for both domestic and international consumption. A budget has been allocated to fund two more full-time village managers to facilitate effective community engagement.

Significantly, the Hahoe experience is informing the Yangdong Management Plan for Visitors by Villagers. It is best practice for sharing expertise and enabling a multiplier effect in serial World Heritage properties.

Significantly, the Hahoe experience is informing the Yangdong Management Plan for Visitors by Villagers. It is best practice for sharing expertise and enabling a multiplier effect in serial World Heritage properties.

Increase in visitor numbers since inscription on the World Heritage List has led to the need for a Visitor Management Plan. It is envisaged that by vividly presenting the authenticity of the site as well as reinforcing its integrity, the plan will encourage visitors to respect the outstanding universal value and practise responsible tourism. It is also realized that to create a better infrastructure for visitors, the village committee should be strengthened.

Yangdong also has a rich legacy of all kinds of human and natural resources. The interpretation of these heritage resources and their accessibility is a priority. In order to minimize possible damage from tourism, concepts and policies for limited visitation are being investigated. Different levels of interpretation are being developed, from virtual to real in format.

The existing village committee is being reconstituted as the World Heritage Yangdong Village Operating Committee. Establishing principles and ensuring a philosophy of empathy, sharing and coexistence are necessities in operating the village committee. Full-time employees deal with general administration,

Jangseung (Totem Pole). World Heritage status and the revitalization of Hahoe intangible heritage has led to the safeguarding of the endangered tradition of totem poles worshipped as sacred and positioned at the entrance to the village as guardians.

site conservation, site management, events, finance, community welfare and community businesses. Priorities are long-term employment and capacity-building, also for government officials.

All projects for the village are classified into general, special or impending. The Operating Committee carries out every project in collaboration with Gyeongju City government, which provides large-scale funding and administrative assistance where necessary. Apart from the government budget, fixed assets acquired and owned by the village are being considered. This will be achieved through fundraising in the short term and introducing the 'National Trust Movement' concept in the long term. According to their characters, the suggested projects will be implemented, divided into three periods, which are 'triggering (2012–13)', 'activating (2014–16)' and 'stable (2017–21)'.

241

20

World Heritage and Chinese diaspora: Kaiping Diaolou and Villages, China

GUO ZHAN[1]

Ancestral homes

The Diaolou, or multi-storeyed defensive village houses of Kaiping, were mainly constructed in the 1920s and 1930s. They display a complex and vibrant fusion of Chinese and Western structural and decorative forms. They reflect the significant role played by émigré Kaiping people in the development of several countries in South Asia, Australasia and North America during the late 19th and early 20th centuries, and the close links between overseas Kaiping and their ancestral homes.

Kaiping Diaolou and Villages is a World Heritage site nestling in the idyllic subtropical countryside of southern China and reflecting the history and traditions of overseas Chinese over the past few centuries with its distinct stamp of cultural exchange between China and different parts of the world. This exceptional architectural form features unique designs perfectly integrating indigenous rural traditions with foreign cultures in such diverse dimensions as architectural planning, land use and landscape design. The buildings are also generally recognized as an outstanding example of harmonious co-existence between man and nature in special natural conditions and historical contexts.

The four selected groups of Diaolou in their landscape represent some 1,800 remaining tower houses still surviving in their village settings, reflecting the culmination of almost five centuries of tower-house building and the strong links remaining between Kaiping and the Chinese diaspora. The Kaiping

> This exceptional architectural form features unique designs perfectly integrating indigenous rural traditions with foreign cultures.

[1] Text provided by the Bureau of Cultural Heritage of Kaiping Municipality and revised by Guo Zhan, Vice President, International Council on Monuments and Sites (ICOMOS).

Thousands of students taking part in the 'Love Our Diaolou Buildings' painting competition (2006). The conservation of the Kaiping Diaolou and Villages World Heritage site has raised awareness of a number of issues in safeguarding its outstanding universal value.

Diaolou site includes four building complexes, Yinglong Building and Sanmenli Village, Zili Village and Fang Clan Building, Majianglong Village and Jinjiangli Village. Its core zone covers an area totalling 372 ha and is inhabited by 283 households with 831 residents. Its buffer zone of 2,738 ha has around 1,000 households with 3,113 residents. Diaolou buildings and natural villages are the most basic units of this World Heritage site, each of them featuring a traditional layout and inhabited by people sharing the same family name and close kinship.

Kaiping development

Kaiping Municipal Government is responsible for the protection, management and monitoring of this heritage site in accordance with the laws and regulations of China. A tourism company, founded by the local government, and local village organizations are jointly responsible for tourism management pertaining to the site.

At the time of the inscription on the World Heritage List in 2007, major threats facing the site included the pressures of modern road and industrial construction as well as unplanned tourist development projects that were imposed on the environment and landscape; impacts of modernization on traditional production and lifestyle, dwindling young population, ageing and disrepair of Diaolou buildings which were built with steel and concrete, the traditional materials and techniques of the early 20th century, loss of traditional artisans and skills, and more.

Most Diaolou buildings were built one century ago or in even earlier periods. Frequent typhoon and flood attacks and lightning strikes have led to damaged and leaking floors, peeling and fractured walls, rotten and broken wooden staircases, doors and screens, damaged sculptures and rusted and eroded ironworks.

In addition to government inspections, preservation of Diaolou buildings is now monitored primarily through regular observation and timely reporting by local residents. More than 90 per cent of artisans and workers repairing and restoring Diaolou buildings are local villagers highly skilled in traditional techniques. About 85 per cent of preserved Diaolou buildings belong to local

243

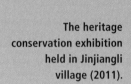

The heritage conservation exhibition held in Jinjiangli village (2011).

More than 90 per cent of artisans and workers repairing and restoring Diaolou buildings are local villagers highly skilled in traditional techniques.

villagers or overseas Chinese. Efforts to preserve their authenticity and integrity in their original state and appreciation of their outstanding universal value have won support and cooperation from all the owners.

Most of these Diaolou buildings are now left empty and locked, and their owners are third- or fourth-generation descendants permanently resident overseas. One of the first significant steps that the local government took for the sustainable development of the site is the design of a system of trusteeship to take care of these Diaolou buildings and encourage their absentee owners to sign contracts of entrusted repair and management with the Management Office for a validity period of thirty to fifty years at no cost to the owner. So far, forty-five Diaolou buildings under government custody are well preserved. For those yet to be entrusted to the government, more than 85 per cent are taken care of through private trusteeship by owners' relatives, friends or local villagers.

Governance

The trusteeship model is driven by the local government which is closer to the stakeholder communities, both owners and residents. After Kaiping Diaolou was designated as a World Heritage site, Kaiping Municipal Bureau of Cultural

Conference for the adoption of the Kaiping Diaolou buildings (2011). The adoption of Diaolou is an innovative approach where an adopter donates only 100,000–300,000 RMB yuan to keep a building for five years.

Heritage was established in early 2008, the first of its kind in Guangdong province, considering that Kaiping is at county level. In May 2009, Kaiping Municipal Center of World Heritage Site Management was set up. In addition, a number of regulations and rules have been enacted in recent years, including the Regulations on Protection and Management of Kaiping Diaolou of Guangdong Province, the Regulations on Protection and Management of Diaolou Buildings and Villages of Kaiping Municipality, the Interim Rules on the Protection and Management of the World Cultural Heritage Site of Kaiping Diaolou and Villages and the Master Plan for Tourism Development of Kaiping Diaolou. Village Committees where Diaolou buildings are located have drafted and adopted Village Rules.

Kaiping World Heritage Property Management Plan has been drawn up by Beijing University under the auspices of the People's Government of Kaiping City, and has been implemented since 2005. The objectives of the plan cover the Diaolou, the villages and their contextual setting. Protective measures are to be put in place for all aspects of the landscape: the spatial layout of the villages, the buildings, rice cultivation, and other agricultural practices, the environment and local customs.

The implementation of the Management Plan is through the Kaiping Protection and Management Office of Diaolou and Villages established in 2000. It has fifteen full-time staff, of which 80 per cent are degree holders in the fields of history, architecture, conservation and maintenance. There are

In addition to government inspections, preservation of Diaolou buildings is now monitored primarily through regular observation and timely reporting by local residents. Here the villager of the Sanmenli Village, local craftsman Mr Fang receiving guidance from professional engineers and technicians.

300 Diaolou keepers working in the villages, all of whom received training before starting work. Allied to this office is the Kaiping Diaolou Research Department, established in 2004 to undertake research on the background history and culture of overseas Chinese and to promote the culture of Diaolou and their villages. The Kaiping Protection and Management Office of Diaolou and Villages is fully integrated into, and has support from, the national protection system through the State Cultural Relics Bureau, the provincial government through the Cultural Bureau of Guangdong Province, and the city level through the Kaiping Cultural Bureau. It also works closely with Management Offices established at village level who appoint Diaolou Protectors and Security Personnel.

Since 1983, the Cultural Bureau of Kaiping has prepared detailed surveys of all Diaolou and their state of conservation. In addition, environmental, economic and population data have been collected. Not all village buildings have been surveyed nor the overall cultural landscape patterns. The Kaiping Protection and Management Office of Diaolou and Villages, established in October 2000, is a well-articulated conservation unit. There is a good understanding of what is envisaged in terms of a heritage-based management system. The State Party is clear that management will be a key factor in the future, particularly in view of the number of absentee owners and the likelihood of increased tourism. The procedures put in place – and outlined above – are considered by ICOMOS to be adequate to address the needs of building conservation and to encourage sustainable development of the wider landscape.

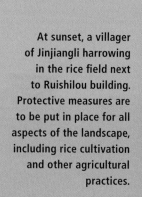

At sunset, a villager of Jinjiangli harrowing in the rice field next to Ruishilou building. Protective measures are to be put in place for all aspects of the landscape, including rice cultivation and other agricultural practices.

The Management Plan is visionary, and based on a well-thought-out analysis of threats and opportunities. It sets out ongoing processes to deal with sustaining the villages as living places that reflect local cultural traditions, and is underpinned by a staffing and consultation structure that is already in place and adequately funded.

Community benefits

A Tourism Management Plan includes details of selling farm produce, serving local farm food and making use of under-used traditional buildings. It is also important to note that a share of tourist income goes to local residents.

Since January 2000, funding has been provided by the People's Government of Kaiping City, the provincial and central governments. Between 2001 and 2005, the government invested US$8,456,800 to stabilize the heritage elements and build appropriate infrastructure. The development of the site between 2005 and 2010 is US$2,416,000. Overseas Chinese contribute to a Kaiping Diaolou Protection Fund which stands at over US$2,000,000. This is managed by the People's Government of Kaiping City.

A designated Tourism Management Plan regulates the way tourism is approached for the overall collection of Diaolou. It is significant that this plan includes details of selling farm produce, serving local farm food and making use of under-used traditional buildings. It is also important to note that a share of tourist income goes to local residents.

A villager of Sanmenli drying paddy near a pond.

248

The Cultural Heritage Bureau of Kaiping Government has organized training programmes to build the capacity of the local conservation professionals in conjunction with the 18 April International Day of Monuments and Sites, the 18 May International Museum Day, the Month of Diaolou Protection, the National Cultural Heritage Day and the Anniversary of the Successful Inscription of Diaolou on the World Heritage List. Consequently, local villagers' knowledge of cultural heritage has increased and their ability to manage cultural heritage enhanced. The local government has made efforts to encourage villagers to continue traditional lifestyles and folk activities.

Capacity-building programmes cater for civil servants who are responsible and influential in maintaining the villages and buildings. Free training is also offered to owners of buildings to maintain the heritage resources. In order to raise the awareness of the next generation and encourage young people in World Heritage conservation, free teaching material is provided to 130,000 students in schools. Workshops and training for local craftsmen to recognize the value of intangible heritage traditions and to record the memories of those involved in the building of the towers are also addressed.

Continued tourism development has on the one hand created more employment opportunities for local villagers, such as janitors, logistic crews, security guards and tour guides, and on the other hand has notably helped to promote

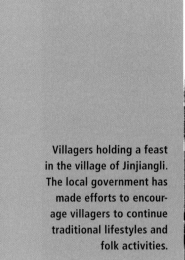

Villagers holding a feast in the village of Jinjiangli. The local government has made efforts to encourage villagers to continue traditional lifestyles and folk activities.

249

economic growth, increase village incomes and the social and moral development of local communities.

Based on the Tourism Development Plan and pilot studies conducted in select villages, several Diaolou have been opened to a restricted number of visitors. The full impact of visitors could be challenging to manage once the properties become better known. The value of the Diaolou lies in the particularly untouched nature of their interiors and current policy is to protect their faded colours as found. Changes in humidity, brought about by large numbers of visitors, and light levels through opening up the towers more than at present, could bring undesirable changes. The carrying capacity of the buildings and ways of restricting numbers at any one time, preventive conservation measures to minimize decay, uncontrolled tourism and development, are being studied.

Increased per-capita income has contributed to a better quality of life for the local villagers and their awareness of the value of the World Heritage property in income generation (Table 1). Most of the residents now live in new houses or apartments and can afford modern home appliances such as cars, refrigerators, TV sets and computers.

It is recorded that local villagers no longer worry about sales of their fruits and crops. One of them, Aunt Guan, said that she can earn up to 300 RMB yuan per day by selling fruit.

Aunt Fang from Zili Village opened a rural-flavour restaurant for the Spring Festival of 2004. She earned a net profit of more than 10,000 RMB yuan within only eight days. Her husband Fang Xiuyong then quit his job and began to run

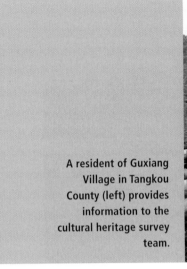

A resident of Guxiang Village in Tangkou County (left) provides information to the cultural heritage survey team.

Table 1 Increase in per-capita income in the four World Heritage site villages (RMB yuan)

Site	2006/07	2010/11	Increase (%)
Majianglong Village	6 000	8 300	-
Luyang Village in Sanmenli	2 843	6 914	143.2
Qiangya Village in Zili	5 206	6 698	28.7
Changle Village in Jinjiangli	4 754	5 995	26.1

the restaurant together with her. As they say, 'Our dishes are quite popular among tourists. The dried vegetable we make can be sold for 40 RMB yuan per kilogram, which can be hardly imagined in other villages.'

Another innovative approach has been to promote the adoption of Diaolou, which is proving effective in encouraging society-wide protection and utilization of these historic buildings. By donating only 100,000–300,000 RMB yuan, an adopter can keep a Diaolou building for five years. The adopter will be honoured with the title of 'Diaolou Protection Ambassador' and offered a certain number of free tickets and can use the adopted Diaolou building to a negotiated extent. So far, twenty-one Diaolou buildings have been adopted and as a result up to 3.3 million RMB yuan raised as repair funds.

Thanks to the cooperation and supervision of local villagers, no roads, buildings and other facilities which are banned by law and planning regulations have ever been built within the World Heritage site or the buffer zone.

Water ponds comprise an integral part of Kaiping Diaolou and Villages. A crescent-shaped water pond is usually located in front of a village, closely integrated with Diaolou buildings, folk houses, ancient banyan trees, bamboo and banana forests and rice fields. The ponds not only play an important role in storing floodwater, discharging pollution and preventing fire accidents, but also function as symbolic 'treasure-gathering basin' designs. Thus they are protected by villagers. Women, primarily responsible for all the above-mentioned work, are major players in preserving and continuing traditional production and lifestyles.

Learning about conservation

The conservation of the Kaiping Diaolou and Villages World Heritage site has raised awareness about a number of issues for safeguarding outstanding universal value. These have informed developments planned for the immediate future, as follows:

- Continuing improvement of conservation strategies for the sites. Revise and amend the Master Plan for Protection of Kaiping Diaolou and Villages based on local experiences and innovation from the first phase of site development.

- Importance of monitoring to inform appropriate responses and timely intervention. Establish the Monitoring Center of the World Heritage Site of Kaiping Municipality and work out the Management Rules for Monitoring of Kaiping Diaolou and Villages.
- Living heritage sites have continuous demands on funding. Diversify the resource base and explore public-private partnerships, including with homeowners, to increase funding for maintenance, repair and restoration of Kaiping Diaolou and Villages.
- Continue cooperation with universities and scientific research institutions both within the country and beyond. Research and further studies in multiple locations helps when studying and conserving the heritage of diaspora communities such as the stakeholders of the Kaiping Diaolou and Villages and cultures of overseas Chinese hometowns.

Conservation through diaspora networking

Kaiping Diaolou and Villages are not only family and economic ties linking clan members and local residents with their overseas relatives, but also physical entities manifesting their common beliefs, identities and historical memories. Thus they are considered part of the physical and spiritual wealth of local communities and present a pattern of harmonious co-existence between man and nature in specific contexts.

Local residents have acquired new knowledge of the landscapes of their hometown with which they had been familiar. They are proud of their hometown and enthusiastic about the protection of its historic monuments.

They have learned the meaning of authenticity, integrity and the basic rules of heritage conservation. The genuine need and care of the local residents as the guardians of Diaolou buildings, the encouragement and support from outside, and appropriate development and improvement, all these together will guarantee the long-term preservation of this heritage site.

21

Role of fisheries and ecosystem-based management: Shiretoko, Japan

YASUKO MIYAZAWA AND MITSUTAKU MAKINO[1]

The 'Shiretoko Approach'

Shiretoko literally means 'the utmost end of the earth' in the local Ainu language. The Shiretoko World Heritage site, in the north-east of Hokkaido, the northernmost island of Japan, consists of the Shiretoko Peninsula and its surrounding marine areas. The distinguishing character of this site is the close link between the terrestrial and marine ecosystems and a number of marine and terrestrial species, including several endangered species.

In Shiretoko, local fishing communities have implemented a wide range of autonomous measures under a co-management framework to maintain responsible and sustainable fisheries. Inscription on the World Heritage List has not led to their exclusion from the area. Instead their activities are placed at the core of the management scheme to sustain ecosystem structure and function. What is significant is that the fisheries co-management was expanded to ecosystem-based management to secure the conservation of this outstanding ecosystem. This collaboration with an integrated focus is called the 'Shiretoko Approach'.

Shiretoko ecosystems

The Shiretoko World Heritage site is the southernmost limit of seasonal sea ice in the northern hemisphere. The rich and complex marine ecosystem is the

---

[1] Ministry of the Environment, Japan, and National Research Institute of Fisheries Science, Fisheries Research Agency. This text is largely derived from Makino, M., Matsuda, H. and Sakurai, Y. 2009. Expanding fisheries co-management to ecosystem-based management: a case in the Shiretoko World Natural Heritage area, Japan. *Marine Policy*, Vol. 33, pp. 207–14.

result of the East Sakhalin cold current running southward towards Shiretoko bringing sea ice; the Soya warm current running south-easterly along the north-east part of Hokkaido towards Shiretoko; and the intermediate cold water derived from the Sea of Okhotsk. In early spring as the sea ice melts phytoplankton blooms, triggered by ice algae, and the intermediate water with its rich nutrient salts form the base of a food chain web that supports the diverse wildlife of the peninsula, including marine mammals, seabirds and commercially important species.

This site is characterized by the interrelationship between the marine and terrestrial ecosystems. A large quantity of anadromous salmonids run up rivers in the peninsula to spawn, and serve as an important source of food for upstream terrestrial species including large mammals such as the brown bear and endangered birds of prey such as Blakiston's fish owl, Steller's sea eagle and the White-tailed eagle, as well as marine mammals. The peninsula is also internationally important as a stopover point for migratory birds.

The marine area is also important for local people, especially for the fisheries sector, which is one of the major industries in the regional economy. Therefore, a coordination system among fisheries bodies and management authorities was required in order to manage the site appropriately and avoid negative impacts on the marine ecosystem from fisheries activities, an issue that was addressed in the lead-up to the nomination and inscription on the World Heritage List.

> This site is characterized by the interrelationship between the marine and terrestrial ecosystems. A large quantity of anadromous salmonids run up rivers in the peninsula to spawn, and serve as an important source of food for upstream terrestrial species.

254

Profile of fisheries

The marine areas around the Shiretoko peninsula are the most productive fisheries in Japan and fishing is one of the main industries. There are 7,706 households with a total of 19,184 people in the towns of Shari and Rausu. In Rausu, more than 40 per cent of the workers over 15 years old are employed by the fisheries. In 2008 Shiretoko fishers caught some 64,000 tons of fish, worth about 24 billion yen (US$313 million). The sustainable use of marine resources is promoted through responsible management, building on the extraordinary productivity of the sea.

The main target species and methods of catch are salmonids by set net, common squid by jigging, and walleye pollock, cod and arabesque greenling by gill net. There are three fisheries cooperative associations in the Shiretoko World Heritage site: Rausu, Shari-daiichi and Utoro. Fish processing and marketing industries are also very active here.

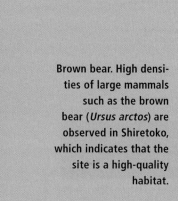

Brown bear. High densities of large mammals such as the brown bear (*Ursus arctos*) are observed in Shiretoko, which indicates that the site is a high-quality habitat.

255

Marine conservation and management

The Shiretoko Approach is based on the establishment of new coordinating mechanisms among a wide range of sectors and stakeholders. This system of coordination and cooperation is one of the most important and characteristic measures to be implemented in Shiretoko. Inscription of the site on the World Heritage List and the commitment to safeguarding its outstanding universal value have been the catalysts for bringing together the different players and local fishing communities.

In October 2003, the Shiretoko World Natural Heritage Site Regional Liaison Committee was established. The Committee is composed of a wide range of sectors, ministries and departments in central and local government, fisheries cooperative associations, the tourism sector and NGOs. The Committee members discuss the appropriate management of the site, exchange information and coordinate various stakeholder interests. In short, it is the core arena for policy coordination among all the stakeholders in the conservation and sustainable management of the World Heritage site.

In July 2004, the Shiretoko World Natural Heritage Site Scientific Council was established. This Council is required to provide scientific advice on the formulation of the Management Plan and on research and monitoring activities.

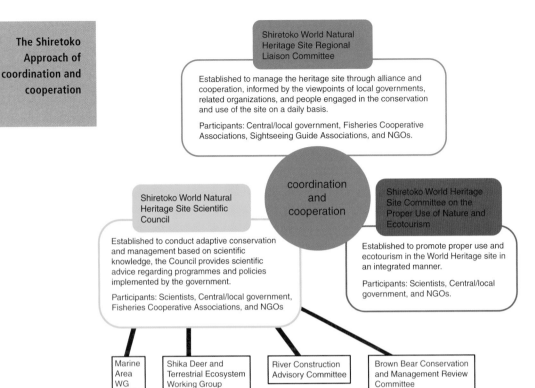

The Shiretoko Approach of coordination and cooperation

Shiretoko World Natural Heritage Site Regional Liaison Committee

Established to manage the heritage site through alliance and cooperation, informed by the viewpoints of local governments, related organizations, and people engaged in the conservation and use of the site on a daily basis.

Participants: Central/local government, Fisheries Cooperative Associations, Sightseeing Guide Associations, and NGOs.

coordination and cooperation

Shiretoko World Natural Heritage Site Scientific Council

Established to conduct adaptive conservation and management based on scientific knowledge, the Council provides scientific advice regarding programmes and policies implemented by the government.

Participants: Scientists, Central/local government, Fisheries Cooperative Associations, and NGOs

Shiretoko World Heritage Site Committee on the Proper Use of Nature and Ecotourism

Established to promote proper use and ecotourism in the World Heritage site in an integrated manner.

Participants: Scientists, Central/local government, and NGOs.

Marine Area WG

Shika Deer and Terrestrial Ecosystem Working Group

River Construction Advisory Committee

Brown Bear Conservation and Management Review Committee

The Council operates through working groups, one of which is the Marine Area Working Group for marine ecosystem management.

These working groups are composed of natural scientists, social scientists, ministries and departments in central and local government, fisheries cooperative associations and NGOs. Their interrelationships have helped to ensure participation, exchange of interests, information and opinions, and the building of consensus between the wide range of users of the ecosystem services, enhancing the legitimacy of the Management Plan. In addition to the above structures for marine conservation and management, there are other organizations to address important issues in Shiretoko World Heritage site. The Sika Deer Working Group, the River Construction Advisory Committee and the Brown Bear Conservation and Management Review Committee are respectively responsible for deer, river and bear conservation and management. The Shiretoko World Natural Heritage Site Committee on the Proper Use of Nature and Ecotourism was founded in 2010, succeeding the Shiretoko National Park Committee for the Review of Proper Use founded in 2001. It conducts research and discussions on visitor management and proper-use rules for tourists. All these organizations represent a wide range of sectors

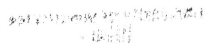

Under the Shiretoko Approach, the local fishers are recognized as an integral part of the ecosystem, and their data and knowledge are officially utilized to monitor the ecosystem and serve as an important foundation for adaptive management of the site.

and deal with topical issues to enable the appropriate integrated management of the World Heritage site.

In December 2007 the Multiple Use Integrated Marine Management Plan was drawn up by the Marine Area Working Group in order to define management measures to conserve the marine ecosystem, strategies to maintain major species, monitoring methods and policies for marine recreational activities. The main objective of the plan is 'to satisfy both conservation of the marine ecosystem and stable fisheries through the sustainable use of marine living resources in the marine area of the heritage site'.

The fisheries sector participated throughout the Marine Management Plan drafting process. Because the ecosystem is disturbed, unclear and complex, the plan stipulates the introduction of adaptive management as a basic strategy. Under the Shiretoko Approach, the local fishers are recognized as an integral part of the ecosystem, and their data and knowledge are officially utilized to monitor the ecosystem and serve as an important foundation for adaptive management of the site. A future task will be to develop reference points representing the overall status and long-term trends of the ecosystem, to be adaptively referred to in the overall management scheme.

257

Responsive conservation

Walleye pollock is one of the most important target fish in the Shiretoko area, and is also a prey of the Steller (northern) sea lion. Shiretoko fishers catch the Nemuro stock of walleye pollock mainly by gill net. This stock is officially managed by the national government under the total allowable catch system based on the Law Concerning the Conservation and Management of Marine Life Resources of 1996.

Despite these efforts, the total annual catch has dropped drastically since 1990. It was around 100,000 tons in the late 1980s, but in 2006 it dropped to only 9,200 tons. To address the decline in catch, local fishers and researchers have cooperatively introduced additional autonomous management measures. Local fishers compile data on catch size, time, area, body size, maturity and so on, which are then sent to the prefectural research station for analysis. The results are returned to the fishers and management measures are discussed and implemented.

Gill-net fishers divided the fisheries ground into thirty-four areas based on their local knowledge and experience. They decided to leave seven of these areas protected to conserve resources. These protected areas include a portion of the spawning ground of walleye pollock. The protected areas are

Extermination of invasive alien species bull thistle.

re-examined every year on the basis of the previous year's performance and scientific advice from the local research station. In addition, six areas were designated as protected areas after the World Heritage nomination.

The reduction of fishing capacity was also addressed and implemented autonomously to conserve resources. There were 193 gill-net vessels in the late 1980s. Local fishers have disposed of more than half of their vessels since 1996 to reduce fishing capacity in accordance with stock status. As compensation for the disposal, about 1.1 billion yen were jointly funded by the remaining fishers and the fisheries cooperative associations. Government bore the interest costs. In 2002, fishers introduced a joint system to reduce fishing pressure by 20 per cent and further reduce operation costs, based on the principle that five boats come together to form a group and each boat suspends operation on a rotational basis.

These autonomous measures with feedback and control are recognized as adaptive management, and officially incorporated in the Marine Management Plan. An important future task will be the scientific verification of the validity of these measures.

With regard to Steller sea lions, the Okhotsk and Kuril populations migrate from Russia to Shiretoko in winter. These populations are listed as endangered on the IUCN Red List. The challenge was to prioritize and properly conserve sea lions. Fortunately, the population has been gradually increasing at 1.2 per cent per year since the early 1990s. On the other hand, from the viewpoint of fishers operating in the Shiretoko World Heritage site, the Steller sea lion is a competitor for walleye pollock resources.

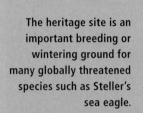

The heritage site is an important breeding or wintering ground for many globally threatened species such as Steller's sea eagle.

Therefore, in order to mitigate the damage, Steller sea lions have been culled each year under the Fisheries Law. The cull limit is set based on the potential biological removal theory. In addition, various efforts are also made to prevent damage to the fishing industry throughout Hokkaido to keep the number of Steller sea lions that need to be cull to appropriate conservation levels.

Marine and terrestrial ecosystem management

As described above, many anadromous salmonids return to rivers in Shiretoko, run upstream to spawn and serve as an important source of food for terrestrial mammals and birds of prey, while contributing to biodiversity and material circulation. On the other hand, salmonids are also an important catch for the fisheries. Under the Fisheries Resource Protection Law of 1951, fishing has been prohibited in all rivers and near the mouths of some rivers in Shiretoko.

To maintain and facilitate the interactions between marine and terrestrial ecosystems, artificial constructions such as dams have been modified since 2006 on scientific advice from the River Construction Advisory Committee. In the four rivers where modifications were completed by 2010, remarkable effects were observed, including increased rates of escapement and spawning bed preparation in the upper reaches above the modified structures.

Clean-up activities are regularly carried out along the coast of the heritage site.

Towards appropriate management

Fishers live directly off local marine ecosystems, so they have accumulated knowledge about the local seas over generations. One of the important lessons from the Shiretoko Approach is the integration of local knowledge and experience along with scientific knowledge in achieving adequate ecosystem-based management. Co-management strategies have many institutional advantages, such as decentralized management, adaptive management, and the use of both local and scientific knowledge. It is important to note that in the Shiretoko Approach, these advantages are recognized and formally incorporated in the Marine Management Plan.

A future task will be to develop a reference point which represents the overall status and long-term trends of the ecosystems, to serve as a benchmark in the management scheme. Progress should be facilitated in the scientific understanding of interrelationships between fisheries operations, indicator species, and ecosystem structure, function and processes.

Under the Shiretoko Approach, a new coordinating system has been established, integrating a wide range of stakeholders from various sectors. This system facilitates the exchange of information and opinions, and strengthens the legitimacy of the Management Plan.

The site is characterized by the interrelationship between the marine and terrestrial ecosystems. Here anadromous salmonids serve as an important source of food for terrestrial species.

Science-based measures implemented in rivers facilitate interactions between marine and terrestrial ecosystems, and procedures to set a limit for the culling of the Steller sea lion have mitigated fishery damage without increasing the risk of extinction.

Building consensus on ecosystem-based management in the fisheries sector, through sharing knowledge and responsibilities for conservation, is an important lesson. Originally, local fishers were apprehensive that World Heritage status would lead to additional regulations for the sole purpose of environmental protection and impact negatively on their livelihood. In order to alleviate these fears, prior to nomination for listing in January 2004, the Ministry of the Environment and Hokkaido Prefecture determined that both the conservation of the ecosystem and stable fisheries would be essential. This assurance was also stipulated as the objective of the Marine Management Plan. In practice, the fisheries sector has participated from the beginning in all the coordinating organizations shown in the figure on page 256 and in the drafting of the Marine Management Plan. Information sessions and meetings have been held several times with local fishing communities. The participation of the fisheries sector, official guarantees and the accountability of administrators were the keys to building consensus.

The participation of the fisheries sector, official guarantees and the accountability of administrators were the keys to building consensus.

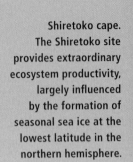

Shiretoko cape. The Shiretoko site provides extraordinary ecosystem productivity, largely influenced by the formation of seasonal sea ice at the lowest latitude in the northern hemisphere.

Community-centred co-management

The Shiretoko Approach is based on the Japanese fisheries community-centred co-management framework, characterized by recognizing fishers as the primary participants in management along with the involvement and support of the broader community. This open-ended participatory approach to a wide range of human needs in the community lends itself to implementation of a balanced mix of biological, social and economic objectives.

In the Shiretoko Approach, the local fishers are an integral component rather than an impediment to the conservation of the 'original ecosystem'. Far from having to be managed or controlled, they are expected to play an indispensable part in ecosystem-based management. In this sense, we hope that the experiences of the Shiretoko Approach could contribute to future ecosystem-based management in other regions where large numbers of small-scale fishers are utilizing a wide range of species under a fisheries co-management regime.

In the report of the UNESCO reactive monitoring mission in February 2008, the mission team applauded the bottom-up approach to management through the involvement of local communities and local stakeholders, and also the way in which scientific knowledge had been effectively applied to the management of the property through the overall Scientific Committee. They also pointed out that these management measures, which we call the Shiretoko Approach, provided an excellent model for the management of natural World Heritage sites elsewhere.

In conclusion, the State Party is to be commended for sustainable efforts to improve the management and protection of this natural World Heritage property. Interagency coordination has been improved through a number of mechanisms as mentioned above and continues to be effective. The governance framework provided through the Regional Liaison Committee and Scientific Council has been enhanced as a workable collaborative vehicle for management.

5 More than the Monumental

Visually imposing World Heritage sites often catch the imagination of the public. Few realize the importance of community engagement that accrues benefits to local people at these sites. This chapter consists of five case studies from different parts of the world that demonstrate local community benefits from World Heritage conservation.

The case study on Memphis and its Necropolis (Egypt) profiles a recent programme to fight poverty in the Dahshur area by fostering sustainable heritage tourism and cultural industries through participatory and community-owned protection and management of Dahshur's cultural and natural resources. The direct beneficiaries are from the five targeted local villages.

World Heritage inscription was a major factor in the conservation of the Wadden Sea (Germany and the Netherlands), particularly to bring about cooperation and coordination and move forward with conservation in a concerted manner. Strategic communication and marketing strategies are being implemented to ensure local community sense of ownership that places the Wadden Sea as part of a larger community of marine World Heritage sites with a leading role along with others such as Australia's Great Barrier Reef and Papahānaumokuākea in the United States.

World Heritage status has boosted the local economy and helped to change negative perceptions of the town at the Blaenavon Industrial Landscape World Heritage site (United Kingdom). Collaboration between the public sector, local traders and community groups has resulted in an exciting events programme intended to assist the local economy and promote the site. The transformations have helped residents to again feel a sense of pride in their historic town and the surrounding heritage landscape, halting the population decline of the past ninety years and contributing to around 100 jobs.

Barnacle goose, Tümlauer Bucht, Germany. In the Wadden Sea up to 6.1 million birds can be present at the same time, and an average of 10–12 million pass through each year.

Prior to the establishment of the Serra da Capivara National Park in north-east Brazil, the region was poor, the infant mortality rate was high and there were no basic public utilities such as water, electricity and communications. Developments subsequent to recognition as a World Heritage site have been astonishing, with all public utilities and services now available.

At Angkor (Cambodia), sustainable development strategies have led to mutual understanding and trust between the authority responsible for the site, park management and residents building a new foundation for long-term collaboration. The development of a new settlement offers future residents a sound infrastructure of roads, agricultural land, an irrigation system, micro-credit facilities, schools, a vocational training centre and a Buddhist monastery, as well as solar energy, windmills to pump water from lakes, organic agriculture and ecotourism networks.

22 Dahshur villages in community development: Memphis and its Necropolis – the Pyramid Fields from Giza to Dahshur, Egypt

AMARESWAR GALLA[1]

Focus on the villages

Dahshur is located 10 km south of Saqqara and 35 km south of Cairo. It is a 3.5 km long field of pyramids forming part of the World Heritage site of Memphis and its Necropolis – the Pyramid Fields from Giza to Dahshur, inscribed on the World Heritage List in 1979. The Dahshur area has significant growth of population and high unemployment. Moreover, it has also been witnessing a rapid growth in tourist visits. A collaborative programme has been initiated, exclusively focused on the Dahshur villages. The principal goal was to facilitate community development to improve living conditions and promote greater ownership of the area.

The MDG-F[2] Culture and Development Joint Programme, entitled Mobilization of Dahshur World Heritage Site for Community Development, was led by the Ministry of Culture (Supreme Council of Antiquities), Ministry of Tourism, Egyptian Environmental Affairs Agency, Social Fund

---

[1] This text is drafted by the Editor of this volume to profile the case study, which is based on information provided by the MDG-F Joint Programme team within the framework of the UNESCO-led MDG-F Knowledge Management Project for Culture and Development.

[2] The Millennium Development Goals Achievement Fund (MDG-F) is an international fund that was established on 18 December 2006 by the United Nations Development Programme (UNDP) and the Government of Spain to accelerate progress in achieving the MDGs and to support the United Nations (UN) Reform through inter-agency interventions at the country level. The MDG-F finances 128 Joint Programmes implemented across forty-nine different countries in eight thematic areas of intervention, including Culture and Development.

Up to 500 children from Dahshur participated in a heritage awareness campaign organized by UNESCO in cooperation with the International Labour Organization in April 2011.

The overall objective of the programme is to fight poverty in the Dahshur area by fostering sustainable heritage tourism and cultural industries through participatory and community-owned processes to protect and manage Dahshur's cultural and natural resources. The direct beneficiaries are approximately 4,900 individuals from the five targeted villages of Dahshur, 39 per cent of whom are women.

for Development and the Industrial Modernization Centre. The international partners were UNDP, UNESCO, UNIDO, UNWTO and ILO. The project period was four years from April 2009. The designated donor support of US$3.1 million came from the Spanish MDG Fund. The targeted area is the Dahshur component of the World Heritage site and the five surrounding villages.

The overall objective is to fight poverty in the Dahshur area by fostering sustainable heritage tourism and cultural industries through participatory and community-owned processes to protect and manage Dahshur's cultural and natural resources. The direct beneficiaries are approximately 4,900 individuals from the five targeted villages of Dahshur, 39 per cent of whom are women. In addition seventeen institutions, including four government ministries, six associations, two regional/local authorities and five non-governmental organizations, have benefited from the project. It is estimated that the indirect benefits have reached 40,000 individuals, 46 per cent of whom are women. It is a well-monitored demonstration project offering valuable lessons and hence its inclusion in this volume as a case study.

Dahshuri women producing craftwork with palm leaves. Different ways of approaching work and learning have been tried out in which women new to handicrafts have participated.

Local communities in sustainable development

In the project framework, sustainable development meant the preservation of Dahshur's pyramids and ecosystem, namely the seasonal lake's unique plant and animal biodiversity, by enhancing capacities to sustainably manage the site and by promoting a responsible approach to tourism development. In this context, the project ensured community development and employment generation with a special focus on women and youth. It fostered community-based tourism by supporting local entrepreneurship and job opportunities in the creative industries and handicraft sectors. It promoted public awareness about the value of Dahshur for social development, making a particular effort to encourage community-owned and operated small and medium-sized enterprises (SMEs).

In the preservation of Dahshur, institutional strengthening progressed the development of an integrated Master Plan for the Dahshur component of the World Heritage site, finalization of a tourism spatial plan for Dahshur (with the government contributing US$8.3 million for implementation), the drafting of archaeological and environmental plans (with the completion of ecological assessment and water analysis), and the submission of a file for Dahshur Seasonal Lake to be declared a Protected Area. Technical training was provided to forty-five officials of the Inspectorates of the Supreme Council of Antiquities at Dahshur, Saqqara and the Giza Plateau in management and protection of archaeological and World Heritage sites. For the first time, twenty-five members of the Dahshur community were invited to participate in this training.

Improved access to key information about Dahshur was facilitated by drawing up a socio-economic profile, the archaeological mapping of the entire area using Geographic Information Systems, and setting up a database on its antiquities. Community participation in managing Dahshur was ensured through the newly established Local Economic Development Forum.

Job creation concentrated on the handicraft, creative industries and cultural tourism sectors, especially for women. Handicraft production was divided into five sectors: palm tree furniture, basketry products, palm rugs, carpet products, beadwork and embroidery. Training of 275 women (including twenty employed) and fifty-five men in traditional handicraft production led to new product designs and improved market links. Cultural entrepreneurship was fostered through financial and technical support. About 1,761 entrepreneurs were provided with business development services and micro-credits were negotiated to start or upgrade their own business. More than thirty locals were trained through two training of trainers programmes to promote entrepreneurship (Know About Business, and Women Get Ahead), resulting in the capacity-building of 300 locals across five villages. Sustainable community-based tourism was also fostered. Over 3,243 locals were trained in Customer Care in English, and in Small and Medium-Sized Enterprise Development in the tourism sector. It is also important to note that awareness-raising of the

Group of Dahshuri artisans producing furniture with palm trees. With the use of local raw materials such as palm branches and leaves, new jobs have been created, generating local revenue.

tourist value of the World Heritage site was facilitated for around 360 local community members.

Project management and governance

The main focus of the project was on fostering community development in the Dahshur area. Highly participatory interventions were undertaken at grassroots level to consolidate community ownership through a bottom-up approach. International, national and local ownership was also ensured by involving a variety of stakeholders in the project.

The Local Economic Development Forum ensured the participation and ownership of the project by the Dahshur community. As a result, local stakeholders, community leaders and representatives of disadvantaged groups (women and youth) were provided with a platform to actively participate in the management and coordination of activities. The use of participatory methodologies such as interactive workshops for training locals in handicraft production were conducted and new product designs were produced based on the recommendations of the stakeholders and community members involved.

271

Network analysis shows that three main categories of actors were involved in the project: the Supreme Council of Antiquities (now the Ministry of Antiquities), UN agencies and NGOs. Various public cultural institutions

Community member from Dahshur village weaving. Training 55 men and 275 women in traditional crafts production has led to new product designs and improved marketing links.

The pyramid of Mycerinus is the smallest of the three pyramids of the Giza Plateau, and part of the Memphis and its Necropolis World Heritage site.

272

Local and regional NGOs were strongly involved in implementing training and workshops. The involvement of local NGOs is very relevant given their capacity to spread knowledge and skills among community members and the fact that they may assist national and local government in designing and implementing future policies, programmes and projects.

closely collaborated with five UN agencies to implement the project activities. Government ownership was consolidated through the active participation of relevant ministries (Tourism, Culture and Environment) in developing strategic management plans for the elaboration of the Dahshur Master Plan. The government's decision to allocate US$8.3 million for the implementation of the Dahshur Spatial Tourism Plan developed by the project is a significant testimony of the extent of government ownership. Ownership among the two participating national agencies (Social Fund for Development and Industrial Modernization Centre) was also ensured through their central role in coordinating technical training.

Local and regional NGOs were strongly involved in implementing training and workshops. The involvement of local NGOs is very relevant given their capacity to spread knowledge and skills among community members and the fact that they may assist national and local government in designing and implementing future policies, programmes and projects.

The project benefits have been spread across the Dahshur community, directly benefiting 4,900 people and indirectly advantaging the 40,000 inhabitants of the surrounding five villages.

Consistent with the aim of promoting broad-based economic activities, the private sector, although not involved in the project design, inception and implementation, particularly benefited from its activities. Indeed, a small unit composed of six local staff was established to provide Business Development Services and micro-credits to the community, as a result benefiting twenty cultural industries and twenty artisans, and enabling over 100 SMEs to be created or upgraded. Moreover, 330 new entrepreneurs have taken part in market-oriented workshops and training programmes in the handicraft sector.

Special attention was paid to supporting local women, who represent 45 per cent of beneficiaries, as evidenced by the design of specific capacity-building activities for women in both handicraft production and entrepreneurism (such as the Women Get Ahead training programme offered to over 100 women). As a result, women are now producing and selling quality handicraft products to local NGOs, taking part in cultural ventures from the five villages and earning an income for the first time.

Finally, state and local authorities have also greatly benefited from the project, in particular the three relevant ministries and their field offices.

Community and ecosystem benefits

273

Various indicators suggesting strong sustainability of the project are worth noting. In particular, the above-mentioned Local Economic Development Forum has successfully been registered as an NGO. It will continue to function after

Dashur is one component of the Memphis and its Necropolis World Heritage site, which also includes the Giza Plateau with its iconic Sphinx and the pyramid of Cephren.

the project along with the newly established micro-financing revolving fund, thereby ensuring organized social action at community level and sustainable access to local finance for entrepreneurs and SMEs. Furthermore, local NGOs are prepared to replicate their training for capacity-building of other groups at community level. Finally, the strategic plans being developed by the project to ensure the conservation and protection of the Dahshur component of the World Heritage site will be integrated into the Management Plan under preparation, ensuring further government investment in protecting the famous pyramids and their ecosystem.

The Dahshur project has demonstrated benefits directly to the local communities in several ways. Participation of community members in cultural tourism has been facilitated through appropriate skills training to build on traditional craft skills. Given the extraction of local raw materials, the focus has been not only on economic development but also environmental sustainability. Training has been given to 200 people, including 140 women, to produce and sell crafts to local NGOs. A primary stakeholder community of artisans and signature products has resulted from the use of local raw materials such as palm branches and leaves. New jobs have been created, generating local revenue.

SMEs have been set up through micro-credit and appropriate skilling in marketing and entrepreneurship based on local knowledge and nuances. What is significant is that an operational revolving fund ensures the viability of the project. It also draws in support from other villagers.

Convincing the local rural community about the benefits of a micro-credit and entrepreneurship programme was a critical step in the process of sustainable economic development. The project identified ninety suitable beneficiaries, about 40 per cent of whom are women. These beneficiaries received business development skilling, micro-credit loans and assistance in establishing or upgrading their businesses. Capacity-building included marketing, entrepreneurship and business development services, together with participation in field assessments, surveys and meetings. The businesses belonged to local community members and a range of participatory activities ensured local ownership. The multiplier effect is evident through local NGO enthusiasm for introducing similar programmes to other groups in the coming months.

Women are the most economically disadvantaged group in the village of Dahshur. The programme aimed at promoting traditional handicraft production among unemployed women. In addition to empowering women, the programme's multiple objectives included reducing poverty, promoting tourism, improving environmental quality and promoting cultural values.

Convincing the local rural community about the benefits of a micro-credit and entrepreneurship programme was a critical step in the process of sustainable economic development. The project identified ninety suitable beneficiaries, about 40 per cent of whom are women.

Seasonal lake with Dahshur pyramids in the background. The process for Dahshur Lake to be declared a Nationally Protected Area is being initiated.

Several challenges have been overcome; new ways of approaching work and learning, finding appropriate venues and expertise, including trainers. A range of participatory methods helped, including rigorous field assessments, cultural mapping and gatherings. Twenty women new to handicraft development participated. Coordination and capacity-building was delivered through a local NGO with a considerable track record in craft production.

The craftwork was then displayed for greater visibility in fairs and visitor centres, and sold in bazaars and markets. It is envisaged that sales in the tourism industry will further local capacities for handicrafts production.

Intensive capacity-building sessions of six months targeting Dahshur traditional crafts were organized by local NGOs. This enabled them to gain their livelihood from producing and selling crafts made from reeds that are quality assured. Recycling of reeds from agricultural waste in the handicraft production has ensured environmental benefits.

It was assessed that the project resulted in 4,900 direct beneficiaries and approximately 40,000 indirect beneficiaries, of whom women constituted respectively 39 per cent and 46 per cent.

Lessons learned

The Egyptian experience has generated significant innovation and knowledge about the role of culture for development in Egypt. Many lessons have been learned, both in the field of culture and development, and in relation to concrete modalities for implementing the UN Reform. Such lessons, which were identified by the Joint Programme team in Egypt, mainly by filling out the MDG-F questionnaire on Culture and Development elaborated by UNESCO, relate both to the executive, operational and financial processes and to the technical aspects of the project.

Appropriate design In projects with strong community involvement, it is critical to allocate adequate time, resources and efforts during the design phase. This is essential for inputs from target communities so that there is local buy-in through addressing relevant needs and priorities. The first voice and ownership of the community members is ensured.

Collaborative implementation Collaborative project development ensures progressive implementation. Cooperation and coordination are facilitated through effective communication between all local, provincial, national and international agencies and partners. There is a mutuality of benefits and shared expertise and leadership for the sustainability of projects. Collaboration is also valuable for benefiting from the range of competencies and expertise offered by the UN system. There are economies of scale in working together where the valuable resources are used for the greatest possible contribution and avoiding duplication, waste and reinventing the wheel.

Viability and sustainability Projects with demonstrated success in benefit-sharing have a higher profile, making them attractive to both stakeholders and funders. This could also lead to a multiplier effect with project expertise informing new initiatives within the target community and beyond.

Valuing culture Valuing successful demonstration projects could be promoted to assess the place of culture in development. For example, as evident from the application of contingency valuation and choice modelling in environmental economics, the cultural economics of outstanding universal value could be further promoted among all stakeholders.

Diversity dividend Inclusive projects facilitate affirmative measures that lead to more equitable benefit sharing and the reduction of inequalities. Women and child poverty alleviation is addressed as an integral part of community development.

Spain's First Vice President Maria Teresa Fernandez de la Vega surrounded by Egyptian children during a short stopover in the village of Dahshur (April 2009). Spain played a determinant role in the creation of the international Millennium Development Goals Achievement Fund (MDG-F).

Beyond the project

277

The success of the project has led to the Government of Egypt's decision to allocate US$8.3 million to the overall programme to capitalize on its promising results after seeing the success achieved on the ground. Economic growth of rural communities through targeted employment-generation activities outside of agriculture in the areas of tourism, cultural industries and handicraft production as a result of local market-oriented training, business development services and micro-financing contributes to diversifying the resource base for project development.

Following on from the benefits of the project, sixty locals have found jobs in garment-making. Fourteen capacity-building workshops benefiting 300 locals from the five villages around Dahshur have contributed to fostering an entrepreneurial culture to harness the development potential of the area while minimizing disturbance to local communities. The Local Economic Development Forum has allowed local stakeholders, community leaders and representatives of disadvantaged groups (women and youth) to engage in organized social action.

Women have been empowered through priority capacity-building activities (over 100 women are now earning an income for the first time), financial support to start their own enterprises, and strong participation in the Local Economic Development Forum where they represent over 30 per cent of community members.

Environmentally sound and responsible tourism is supported and promoted in Dahshur by preparing an Environmental Management Plan for the area and by initiating the process for Dahshur Lake to be declared a Nationally Protected Area. There is also a view to recommend the inclusion of Dahshur Lake and the Palm Grove in the buffer zone of the World Heritage property. Training has been given to 220 locals to produce and sell craftwork using raw materials such as palm tree branches, leaves and reeds, thereby addressing environmental degradation through eco-friendly methods of production.

Conclusion

The Dahshur programme illustrates the usefulness of targeted demonstrated projects where the World Heritage property could become a catalyst for community development. Outstanding universal value becomes an attraction for investment in conservation. Accrued benefits ensure that the local communities take ownership not only of the project process but the source or key locality of the project that is the World Heritage site.

It is heartening to see that the Dahshur programme is building on its success. The proposal for designing a second phase for the Mobilization of the Dahshur World Heritage Site for Community Development received overwhelming support at the meeting of the programme's National Steering Committee at the end of October 2011,[3] attended by representatives of the implementing partners from national and international organizations.

In late 2011, Index Trade Fair in Dubai invited the Turatheyat NGO and showcased the work and handicrafts of Dahshur women artisans. The theme was environmentally friendly products inspired by Pharaonic art in Memphis. The fair provided an entry into interior design for the women. The exhibition is valued for its unique products and the goal of locating creativity in positive transformations in people's lives which in turn will alleviate poverty.

[3] *Dahshur Newsletter*, 2011, Issue 5, October/November.

23

Sustainable development in a Dutch-German World Heritage site: The Wadden Sea, Germany and the Netherlands

BARBARA ENGELS AND CAROL WESTRIK[1]

Cross-border partnerships

The Wadden Sea covers an area along the North Sea coast of Denmark, Germany and the Netherlands. It is a 'large temperate, relatively flat coastal wetland environment, formed by the intricate interactions between physical and biological factors that have given rise to a multitude of transitional habitats with tidal channels, sandy shoals, sea-grass meadows, mussel beds, sandbars, mud flats, salt marshes, estuaries, beaches and dunes.... The [site] is home to numerous plant and animal species, including marine mammals such as the harbour seal, grey seal and harbour porpoise.... The site is one of the last remaining natural, large-scale, intertidal ecosystems where natural processes continue to function largely undisturbed'.[2]

In 1978 the first Governmental Conference on the protection of the Wadden Sea took place between the three countries. The *Joint Declaration on the Protection of the Wadden Sea* was signed in 1982. It formed the basis of the World Heritage nomination. In 2009 the Dutch and German part of 973,562 ha/9,735.6 km^2 was inscribed on the World Heritage List for its uninterrupted system of intertidal sand and mud flats, the largest in the world. The natural processes are undisturbed throughout most of the area. The nomination file states that the 'underlying approach of the conservation and sustainable use of the nominated property is an ecosystem approach'. All habitats which

---

[1] Barbara Engels, World Heritage expert, Bonn, Germany and Carol Westrik, Heritage Consultant, The Netherlands.
[2] http://whc.unesco.org/en/list/1314.

belong to the Wadden Sea are encompassed by the conservation regime in order to protect the ecological processes that are fundamental to the conservation of the system and its flora and fauna.

The World Heritage site borders three provinces in the Netherlands and three Federal States in Germany. Although the World Heritage site is the intertidal sand and mud flats environment, about 3.5 million people live along the coastline or on the nearby islands (Prognos AG, 2004). Therefore, the involvement and support of local communities in the World Heritage nomination process was crucial. This was a long process, with ups and downs. With the inscription of the Wadden Sea on the World Heritage List public awareness and involvement of local communities has been growing. This case study reflects on this process in more detail.

The Wadden Sea is a major source of income, for example from fishery and tourism. It is also an essential recreational area that is densely populated. It is one of the most important areas for migrating birds in the world and is connected to a network of other key sites for migratory birds. It is possible to find up to 6.1 million birds at the same time in the area and an average of 10–12 million pass through it each year. The principal challenge is the reconciliation of seemingly opposing goals of tourism and conservation so as to contribute towards the sustainable future of the area.

The road towards nomination

Denmark, Germany and the Netherlands decided to work towards nomination for World Heritage status as early as 1991. However, it was in 2005 that Germany and the Netherlands planned a joint nomination. It was at the time politically not feasible for Denmark to join the nomination, but it has always been the desire to complete the World Heritage site with the Danish part. The basis for conservation of the Wadden Sea is the Trilateral Wadden Sea Cooperation including Denmark. At the 11th Trilateral Governmental Conference on the Protection of the Wadden Sea in March 2010, the ministers agreed 'to start in the forthcoming period a possible nomination of the Danish Wadden Sea' (CWSS, 2010) as an extension of the inscribed World Heritage property.

The Wadden Sea Plan (WSP), established in 1997 and renewed in 2010, is the coordinated management plan for the Wadden Sea Area, thus not only for the World Heritage site but also for the Danish Wadden Area. It is a combination of the national and regional rules and regulations aiming to secure coordinated management of the Wadden Sea. It is also a policy plan, which

Terschelling, Netherlands. The Wadden Sea is a large, temperate, relatively flat coastal wetland environment, formed by the intricate interactions between physical and biological factors.

means that it includes the vision, shared principles, targets and policies and management measures, together with actions to be taken with roles and responsibilities. In the implementation of the WSP the Trilateral Monitoring and Assessment Programme has become an effective tool (CWSS, 2008, pp. 133–34).

Although it is a natural World Heritage site with hardly any inhabitants – only three people live within the property – ensuring the support of the surrounding local communities was vital for the nomination (CWSS, 2008, pp. 123). IUCN noted in their evaluation of the site or the designated area that even though 'most activity takes place outside the nominated property, all activities are intimately linked to its values, and tourism and recreational activities are a substantial part of the public use and regional economic development in the nominated property' (World Heritage Committee, 2009b, pp. 17–28).

One of the first steps in promoting the involvement of stakeholder communities was to inform them about the World Heritage Convention and the purpose for nominating the Wadden Sea. This involved organizing information meetings at the various municipalities that border the site. In the Netherlands, for example, an information meeting was held on the island of Texel. This attracted groups both in favour and against World Heritage nomination. It was essential to get the message of the Convention across and to make it clear that this was an international Convention, not a national title that would involve more rules and regulations. In Germany the information and consultation was realized via the Boards of Trustees and Advisory Boards of the national parks which are composed of representatives of local and regional governments, as well regional stakeholders representing commercial, recreational and environmental interests. The added value of the World Heritage status was often questioned. There was apprehension about limitations on future developments and activities. These concerns were already identified as significant in a feasibility study on the potential nomination of the Wadden Sea in 2000 (Burbridge, 2000, p. 22).

In order to ensure effective communication with the local communities, a special website was designed for possible World Heritage nomination of the area. People could leave messages on this website and ask questions. Information was also disseminated, for example through DVDs, flyers and brochures, about the purpose of the World Heritage Convention and possible implications of World Heritage status for the Wadden Sea area.

281

Poster of the Joint Campaign 2010–2011. The objective of the campaign was to follow up communication and awareness activities, and to further promote and safeguard protection. ('There is a place – where heaven and earth share the same stage; Our World natural heritage Wadden Sea – experience and protect a natural wonder').

In order to ensure effective communication with the local communities, a special website was designed for possible World Heritage nomination of the area. People could leave messages on this website and ask questions. Information was also disseminated, for example through DVDs, flyers and brochures, about the purpose of the World Heritage Convention and possible implications of World Heritage status for the Wadden Sea area. An important moment in obtaining support from the local communities was the commitment by the Dutch Minister of Agriculture, Nature and Food Quality that World Heritage status would not result in additional regulations or restrictions within the area. All these efforts ensured the commitment of the local communities. It was also made clear that it is essential to keep the lines of communication open at all times.

The website continues to be an important medium of communication and interaction between all stakeholders. There is a space on the website where people are invited to leave their own stories; The Wadden Sea Stories. The Wadden Sea also has its own Facebook page.[3] Education is an important component in the conservation of the World Heritage property. There is a broad range of educational activities including various excursions, by foot, bike or

[3] http://www.facebook.com/WaddenSea.WorldHeritage.

Local communities celebrating the inscription of the Wadden Sea. Since the inscription of the Wadden Sea on the World Heritage List, public awareness and involvement of local communities has been growing.

283

boat. The objective is to increase the knowledge of the people concerning the 'preservation and recovery of natural and cultural values (landscapes, eco-system, and species) in the Wadden region and the North Sea' (CWSS, 2008, pp. 143). After inscription on the World Heritage List, an 'Action Plan for the communication of the Wadden Sea World Heritage' was developed in coop-eration with the stakeholders.[4] Cooperation is further strengthened by regular stakeholder workshops bringing together regional and local actors from both the Netherlands and Germany to exchange experiences and discuss relevant issues linked to the property.

In order to integrate World Heritage-related issues in the various areas of the Trilateral Wadden Sea Cooperation (e.g. international cooperation, research, communication), it has appointed a World Heritage Task Group which reports to the Wadden Sea Board, the governing body of the Cooperation. This Task Group prepares all relevant projects and documents relating to World Heritage. Whereas inscription on the World Heritage List has not resulted in any addi-tional institutional funding arrangements, it has triggered various activities and projects directly or indirectly linked to the new status. Additional funding has been made available at different levels (European, federal, regional and

[4] http://www.waddensea-worldheritage.org/nl/news/2009-11-11-action-plan-communication.

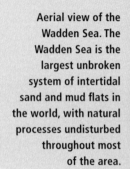

Aerial view of the Wadden Sea. The Wadden Sea is the largest unbroken system of intertidal sand and mud flats in the world, with natural processes undisturbed throughout most of the area.

local) following applications by the stakeholders. One example is the German Federal investment programme for national World Heritage sites where 2.08 million euros have been granted to finance investments for information centres or regional tourism concepts in the Wadden Sea region.[5] The key to adequate funding is access to the right funding programme and, last but not least, a consistent planning and programming of the activities.

Sustainable tourism

Tourism and recreational activities are a substantial part of the public experience of the Wadden Sea. They are unique opportunities to experience the World Heritage values. With approximately 10 million tourists staying overnight, and 30–40 million day visitors coming to the site every year, tourism contributes substantially to the local and regional economy. For example, the national park in Lower Saxony received 20 million visitors in 2006/7, leading to a gross turnover of 115.8 million euros which is equivalent to 3,360 jobs (Job, 2009, p. 139). On the other hand, recreational activities and tourism could also have a negative impact on the values of the Wadden Sea. Although most touristic activities take place outside the World Heritage site, they are intimately linked to the conservation area. A wider approach is therefore

[5] http://www.welterbeprogramm.de/cln_031/nn_613348/sid_414E87DCCB0D0A282094D9B91
B8B0B55/nsc_true/INUW/DE/Projekte/Kommune/Wattenmeer/wattenmeer__node.html?__
nnn=true.html.

Community support for the World Heritage Wadden Sea. Constructive community engagement, cooperation and collaboration bring together nature and culture in safeguarding the outstanding universal value of the site.

needed to ensure sustainable tourism development. The World Heritage Committee made this very clear when inscribing the property on the World Heritage List in 2009 with its recommendation for an 'overall Tourism Development Strategy for the property that fully considers the integrity and ecological requirements of the property and that provides a consistent approach to tourism operations in the property' (World Heritage Committee, 2009a).

Whereas at regional level various approaches to sustainable tourism have already been developed, an overall sustainable tourism strategy is much harder to achieve. In order to best deal with this challenge, the Trilateral Wadden Sea Cooperation has decided to establish a specialist Task Group Sustainable Tourism Strategy composed of representatives from both nature conservation and tourism. This group is steering the process. With Denmark also being involved in the project, it is ensured that the strategy will fit to a potentially extended World Heritage property. According to the work plan the draft strategy will be presented to the Wadden Sea Board in mid-2012.

Funding was successfully obtained from the EU INTERREG IVB North Sea Region Programme for identifying opportunities and perspectives for sustainable socio-economic development in the Dutch-German-Danish Wadden Sea region from the designation of the Wadden Sea as UNESCO World Heritage. The so-called PROWAD[6] project is led by the Common Wadden Sea Secretariat (CWSS) and includes a broad range of partners: the Dutch Ministry of Economy, Agriculture and Innovation, the Regiocollege Waddengebied, the National Park Administration Lower Saxon Wadden Sea, the National Park Administration Schleswig-Holstein Wadden Sea, the World Wildlife Fund Germany, as well as the Danish Ministry of the Environment – Nature Agency. A participatory approach with relevant local and regional stakeholders has been chosen to develop the strategy. On the basis of an action plan, joint

285

[6] Protect & Prosper: Sustainable Tourism in the Wadden Sea.

Experiencing the
Wadden Sea in
Westerhever Sand,
Germany.

The transnational
sustainable tourism
strategy and action
plan will emphasize
the specific role of
the tourism sector for
nature conservation
and for economic and
social welfare.

projects will be implemented supporting regional development and marketing of high quality tourism products.[7]

PROWAD analyses the potential for sustainable tourism and carries out regional workshops with stakeholders from the nature and tourism sectors, developing a joint vision, aims and principles. The transnational sustainable tourism strategy and action plan will emphasize the specific role of the tourism sector for nature conservation and for economic and social welfare. On the basis of the tourism strategy, PROWAD will develop marketable sustainable tourism offers for various target audiences at national and international levels. Guidance on the inclusion of sustainability elements in tourism offerings will also be prepared. The focus will be on climate and nature-friendly activities underlining the transnational character of the Wadden Sea. With the project design also covering implementation, a good basis for success is established.

Joint World Heritage campaign

The designation of the Wadden Sea as World Heritage in 2009 led to a new phase of Wadden Sea communication, awareness and information activities. The recognition of the outstanding universal value of the Wadden Sea ecosystem was considered to be of new quality and delivered added value for the Trilateral Cooperation and the State Parties. In order to communicate this new quality

[7] http://www.waddensea-secretariat.org/management/prowad.html.

Stakeholder commitment is strengthened by regular workshops bringing together regional and local actors from both the Netherlands and Germany.

287

to a broad number of target groups to enhance the awareness of the Wadden Sea at local, regional and international levels, all stakeholders agreed to develop together with the tourism sector a joint World Heritage site campaign in 2010. The objective of the 2010 Wadden Sea World Heritage campaign was to follow up the communication and awareness activities launched on the occasion of the inscription of the Wadden Sea on the World Heritage List in June 2009 and to further promote and safeguard protection, management and awareness, to support stakeholder cooperation and networking in order to strengthen the common responsibility for the site and to support a regional sustainable development strategy. 'There is a place where heaven and earth share the same stage' was chosen as a slogan for the joint campaign, referring to one of the main features of the Wadden Sea – its tidal flats where the horizon meets the ground. The campaign delivered joint communication material which all partners integrated into their activities to ensure a consistent presentation of the World Heritage property.

The central part of the 2010–2011 campaign was the story hunter events. From the end of April until the end of June 2011, three teams (one from the Netherlands and two from Germany) visited various sites, events, and so on in the regions and collected personal stories relating to the Wadden Sea from both inhabitants and visitors. In total, 360 stories (videos, text, photos) were collected during 34 events in both Germany and the Netherlands. These stories were presented via the World Heritage website and Facebook.[8] Press, regional

---

[8] http://www.waddensea-worldheritage.org/Campaign.226.0.html and http://www.facebook.com/WaddenSea.WorldHeritage.

Tourism and recreational activities are a substantial part of the public experience of the Wadden Sea.

The World Heritage inscription has become a major factor in the conservation of the Wadden Sea, particularly to help steer 'all noses in the same direction', and move forward with conservation in a concerted manner. The site is not just developing a communication strategy but also a marketing strategy, one that places the Wadden Sea as part of a larger community of marine World Heritage sites with a leading role along with others such as the Great Barrier Reef and Papahānaumokuākea.

and nationwide media, including social media channels, were used to promote the campaign to a wider audience. As a result the campaign not only reached an additional coverage in the media but also gave a closer insight into local inhabitants' and visitors' awareness concerning the Wadden Sea, both as an area and as a World Heritage site. This will allow a better tailored approach and cooperation in the future.

The World Heritage inscription has become a major factor in the conservation of the Wadden Sea, particularly to help steer 'all noses in the same direction', and move forward with conservation in a concerted manner. The site is not just developing a communication strategy but also a marketing strategy, one that places the Wadden Sea as part of a larger community of marine World Heritage sites with a leading role along with others such as the Great Barrier Reef (Australia) and Papahānaumokuākea (United States).

More specifically on this international cooperation towards the promotion of sustainable development, the Wadden Sea has agreements with other Marine Park authorities around the world within the World Heritage community, especially those that are points of migration for the birds from the area. One example is the capacity-building cooperation with Banc d'Arguin National Park (Mauritania) where Wadden Sea birds overwinter.

Community engagement safeguarding outstanding universal value

In preparing the nomination of the Wadden Sea as a World Heritage site, the main focus was on joint transnational activities for celebrating the site's outstanding universal value and many years of conservation efforts. Involving the local communities in this process was, and still is, crucial. Stakeholder commitment, both from the local population and the government, is further strengthened by regular workshops bringing together regional and local actors from both the Netherlands and Germany to exchange experiences and discuss relevant issues linked to the World Heritage property. An important aspect of the road towards the nomination was to get across the message of the Convention, what it is, what it does and what it could do – both opportunities and challenges. After inscription the call from the stakeholders for delivering additional benefits to the region became more prominent.

Today, three years after the World Heritage Committee acknowledged the Wadden Sea's outstanding universal value, transnational cooperation is strengthened to safeguard this natural heritage for future generations and to use the World Heritage as a 'place making catalyst' for sustainable development (Rebanks…, 2010, p. 2). The existing management structures were fine tuned following the inscription. Place making is often associated with cultural values only, but the Wadden Sea gives a new meaning to the concept in the way that constructive community engagement, cooperation and collaboration bring together nature and culture in safeguarding the site's values.

24

World Heritage site status – a catalyst for heritage-led sustainable regeneration: Blaenavon Industrial Landscape, United Kingdom

CATHERINE THOMAS[1]

Introduction

The Blaenavon Industrial Landscape World Heritage site is located on the north-eastern rim of the South Wales coalfield in the United Kingdom. Today, it still shows evidence of the extensive coal mining and ironworking that took place during the Industrial Revolution.

In December 2000, a 33-km^2 area of the Blaenavon Industrial Landscape was inscribed as a World Heritage site and recognized for its outstanding universal value. The site met two of the prescribed selection criteria for inscription: (iii) the Blaenavon Landscape constitutes an exceptional illustration in material form of the social and economic structure of 19th-century industry; (iv) the components of the Blaenavon Landscape together make up an outstanding and remarkably complete example of a 19th-century industrial landscape.

The landscape includes the Blaenavon Ironworks, dated around 1789, the best preserved blast furnace of its type and period in the world. The Big Pit: National Coal Museum, dating from the mid 19th century, is a complete coal mine and is the best preserved monument of the great South Wales coalfield. The landscape exhibits numerous historic mineral workings and waste tips. Waymarked footpaths following the tracks of some of the earliest iron railways, lead to the Monmouthshire and Brecon Canal, an internationally significant

[1] Head of Regeneration and Coordinator for Blaenavon Industrial Landscape World Heritage, Torfaen County Borough Council, Wales (United Kingdom).

The historical costume procession held annually on World Heritage day reflects the rich industrial heritage of the site.

waterway that provided the early export route for iron and coal. The town of Blaenavon, which grew following the opening of the ironworks, is the best pre-served iron town in Wales and features many interesting social, religious and educational buildings including St Peter's School, provided by the ironmaster's sister, Sarah Hopkins, for local children in 1816 and the traditional two-storey stone and brick terrace with slate-roofed houses, that provided homes for the ironworkers and colliers.[2]

In summary the reason for the nomination was that:

'The Blaenavon Industrial Landscape presents a large number of individual monuments of outstanding value within the context of a rich and continuous landscape, powerfully evocative of the industrial revolution. It is one of the prime areas in the world where the full social, economic and technological process of industrialisation through iron and coal production can be studied and understood' (DCMS, 1999).

Socio-economic background

The decline of heavy industries in the 20th century had a devastating impact on the town of Blaenavon. The population fell from 12,469 in 1921 to under 6,000 in 2001. Industrial decline posed many challenges and by the end of the 20th century the plight of the town was plain to see. What had become

[2] Blaenavon World Heritage Site Management Plan 2011–2016, Draft Statement of Outstanding Universal Value: Brief Description.

'Party in the Past' at Blaenavon Ironworks. In September 2010, Blaenavon celebrated a decade of World Heritage site status with a spectacular open-air event.

a decayed dormitory town, with its boarded-up buildings, epitomized the negative stereotype of a depressed post-industrial community.

The poor social and economic conditions facing the community of Blaenavon were recognized by local, regional and national government organizations and led to the development of the Blaenavon Heritage and Regeneration Strategy. This provided the impetus for obtaining World Heritage site status as a catalyst for regeneration and the setting up of the management structure to ensure its long-term sustainability, both in terms of conservation and economically and socially.

Management of the World Heritage site

The site is managed by the Blaenavon World Heritage Site Partnership, which brings together a group of public-sector bodies and organizations under the leadership of Torfaen County Borough Council. Their shared vision is:

> 'To protect the cultural landscape so that future generations may understand the outstanding contribution South Wales made to the Industrial Revolution. By the presentation and promotion of the Blaenavon Industrial Landscape, it is intended to increase cultural tourism, provide educational opportunities and change perceptions of the area to assist economic regeneration'(Blaenavon World Heritage Site Management Plan 2011–2016).

Over the past decade the Blaenavon World Heritage Site Partnership has succeeded in delivering a significant, £35 million heritage-led regeneration programme for the benefit of the local community. This investment has resulted in the protection, conservation, upgrading, promotion and presentation of the national monuments of the Big Pit and the ironworks, in addition to renewal and environmental improvements to the historic town and gateways, waymarked walks, the relict landscape and the restoration of two derelict listed

World Heritage status has boosted the local economy and helped to change negative perceptions of the town. The Blaenavon World Heritage Site Partnership aims to develop the area as an internationally recognized visitor destination and the development of cultural tourism has already created or safeguarded around 100 local jobs and should offer more employment opportunities in future.

buildings. These two buildings were the original schools provided for the iron-workers' children. They have been extensively refurbished and are now the Blaenavon World Heritage Centre, providing visitors with both intellectual and physical orientation of the World Heritage site.

Economic activities and benefits

World Heritage status has boosted the local economy and helped to change negative perceptions of the town. The Blaenavon World Heritage Site Partnership aims to develop the area as an internationally recognized visitor destination and the development of cultural tourism has already created or safeguarded around 100 local jobs and should offer more employment opportunities in future.

The Big Pit: National Coal Museum now offers guided underground tours conducted by former coal miners. The mine receives over 160,000 visitors a year. The restored pithead baths feature displays and interactive exhibitions enabling visitors to learn about the history of coal mining, and the people who lived and worked in Welsh mining communities.

Blaenavon Ironworks is the most significant historical monument within the Blaenavon Industrial Landscape World Heritage site. Today visitors can view the extensive remains of the blast furnaces, cast houses and the iconic water balance tower. The ironworks has also been host to many cultural and community events. In September 2010, Blaenavon celebrated a decade of World Heritage site status with a spectacular open-air event. An evening of outstanding musical entertainment, which included the Blaenavon Male Voice Choir and Town Band, was enjoyed by residents and visitors. Schoolchildren participated and performances were given by a popular local drama group. The evening culminated in a tremendous firework display which could be enjoyed from all corners of the town.

Within the town, various new specialist shops have opened since 2000 and stand proudly next to Blaenavon's established traditional businesses. Collaboration between the public sector, the local traders and community groups has resulted in an exciting events programme intended to assist the local economy and promote the World Heritage site. Local produce markets and Christmas fayres have helped to increase footfall in the town centre.

The Blaenavon World Heritage Site Partnership has invested heavily in branding and coordination of its cultural assets to ensure that the town's historical attractions now feel like a combined offer for residents and visitors alike. The result is a town that has taken its image very seriously indeed

Bird walk. Guided walks programmes have proved popular with both locals and visitors alike, engaging people in healthy activities while learning about the World Heritage site landscape.

and can genuinely claim to have a distinct identity (Rebanks …, 2009).

Blaenavon, although very conscious of its proud past, actively embraces the future. It is now home to one of the most advanced public-sector ICT projects in the UK, which has brought new jobs and a skilled workforce to the area. The Shared Resource Service, based next to the area's most historic sites, is providing a range of digital services as part of a 'Digital Valley' vision to make the region one of the most connected in the country. The town, once at the heart of the Industrial Revolution, is now playing a significant part in bringing about a 'digital revolution' in the communities of south-east Wales.

Investment has substantially improved the Gilchrist Thomas Industrial Estate, allowing Blaenavon to attract and retain innovative companies. This has included the upgrading of Doncaster's, a legacy of the iron and steel-working tradition. A local business, SuperRod, manufacturers and distributers of cable-routing tools for the electrical industry, won the Queen's Award for Innovation in 2008 and new developments, such as the Blaenavon Brewery, have enhanced the heritage landscape through sympathetic design.

Innovation and heritage have been encouraged to work hand in hand, offering the town and its hinterland a sustainable future.

Social regeneration through community engagement and local skills

Education and interpretation play an important social role, engaging the community and visitors with their heritage, and promoting learning. For this reason, both aspects have been actively developed since inscription of the site in 2000. Thousands of people, young and old, now enjoy formal and informal educational opportunities at each of the main attractions every year.

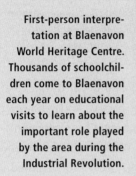

First-person interpretation at Blaenavon World Heritage Centre. Thousands of schoolchildren come to Blaenavon each year on educational visits to learn about the important role played by the area during the Industrial Revolution.

The Blaenavon World Heritage Centre provides a range of educational resources suitable for schoolchildren, covering the national curriculum subjects of history, geography and design & technology. The staff also work with the community on local heritage projects. A local history group has recently received guidance and assistance with research and funding to publish a social history of Blaenavon entitled Funeral to Festival.[3] The book combines the diverse lifestories and memories of group members with archive sources and is enriched by heartwarming illustrations by a talented local artist. Importantly, this is a book produced solely by people who love their heritage and their home town of Blaenavon, rather than by professional historians.

The World Heritage Centre also provides an essential platform for local people to express their heritage. Many of the exhibitions displayed in its main gallery have had local heritage themes and have been developed by community organizations including Blaenavon Male Voice Choir, Blaenavon Art and Artist Association and local photography clubs. The staff have established relationships with small community groups to help tell their stories, and the local people involved submit photographs, memories and documents relating to the area. These memories are being shared online using the Visit Blaenavon website and it is hoped that this will expand over the coming years for future generations to enjoy.

Nestled within the cultural landscape is the Blaenavon Heritage Railway, just a short walk from the Big Pit: National Coal Museum, from which it would

295

[3] Funding provided by the Forgotten Landscapes Project, which is principally supported by the UK Heritage Lottery Fund.

The Welsh Coal mining museum – Big Pit in Blaenavon, South Wales. The area around Blaenavon is evidence of the pre-eminence of South Wales as the world's major producer of iron and coal in the 19th century.

Today the railway has become a key visitor attraction within the World Heritage site and in 2011 around 13,000 visitors enjoyed a 6-km round trip taking in the sights of the industrial landscape. It is entirely managed by local volunteers, many of them having received training in operation and maintenance.

have transported coal down to the coast at Newport in its heyday. Today however the railway has become a key visitor attraction within the World Heritage site and in 2011 around 13,000 visitors enjoyed a 6-km round trip taking in the sights of the industrial landscape. It is entirely managed by local volunteers, many of them having received training in operation and maintenance. These volunteers have extended the railway over the past two years to improve the visitor experience, to include a new station building at Blaenavon High Level, as well as constructing a new branch line linking it with the Big Pit and the town. This is a perfect example of what can be achieved through the enthusiasm of local people and exemplifies how passionately they feel about their heritage.

The Blaenavon World Heritage Site Partnership sees the community and the voluntary sector as having a vital and increasing role to play in the protection, promotion and management of the site's outstanding universal value and believe the best ambassadors for this are found in the local community.

Since achieving World Heritage status, the community has celebrated with an annual World Heritage Day event which is held on the last Saturday of June. The event is organized by the voluntary World Heritage Day Community Group with assistance of the local authority. Each year resident artists work with local schoolchildren to create their costumes for the historical costume procession, with the theme always based on their rich industrial heritage. This has over several years helped to create civic pride in the town, helping to protect it from anti-social behaviour. The day is enjoyed by the whole community, from the schoolchildren to their parents and indeed their grandparents.

Environmental improvements

Through collaboration between the Blaenavon World Heritage Site Partnership, community groups and local businesses, plans to address the dereliction of the town have been successfully implemented. Substantial public- and private-sector investment has been undertaken in Blaenavon over the past decade, some examples of which are:

- Rebranding as a heritage town. At least 75 per cent of the town centre dereliction has been made good, providing accommodation for new trading businesses.
- Refurbishment of a vacant and derelict hotel and its conversion into bed and breakfast accommodation comprising eleven en-suite bedrooms, equipped to Visit Wales standards and restaurant. Due to open summer 2012.
- Renovation of 500 residential and commercial properties.
- Car parks and public amenities have been greatly improved and anti-social behaviour has declined significantly.
- Renovation of two derelict public buildings in particular, the former Council Offices and the former St Peter's School, provides examples of conservation and restoration being used for sustainable purposes within the community. The Council Offices now house the town library, and the school has been 'reborn' as the United Kingdom's first dedicated World Heritage Centre, providing both a community resource and a focal point for understanding and enjoying the World Heritage site.
- A major private housing development has begun, the first for at least fifty years, using a layout and design of high conservation standards within the World Heritage site.

All these improvements have helped residents to again feel a sense of pride in their historic town and the surrounding heritage landscape, halting the population decline of the past ninety years.

> All these improvements have helped residents to again feel a sense of pride in their historic town and the surrounding heritage landscape, halting the population decline of the past ninety years.

297

Sharing Blaenavon's experience and knowledge

Much of the success brought about in Blaenavon has been due to strong and effective partnerships. Major investments in the town have been facilitated by the 'discretionary' nature of the contributions by the various partners. Funding was only allocated by individual partners when they could see the benefits of the project and it fell within their own criteria.[4]

---

4 Criteria: the specific elements of a project a partner can fund through their grant scheme. An example would be one of the partners providing grant funding towards the 'conservation' elements within a larger regeneration project.

The importance of effective coordination in forward thinking, strategic planning and ensuring ongoing personal contact within the Partnership and the local community has been and continues to be crucial to its success. Close contact with the UK Committee of the International Council on Monuments and Sites (ICOMOS UK) and other parties such as the Local Authorities World Heritage Forum (LAWHF) helps to ensure that management is being undertaken in accord with the latest UNESCO guidance and thinking and with 'best practice' elsewhere in the UK and the rest of the world.

The need to find a balance between protecting its heritage and the need for positive change has been crucial to Blaenavon's advancement. Before World Heritage status was achieved, the 'heritage' had suffered through neglect due to lack of awareness and investment, as the area had become one of the most socially and economically depressed in Wales and indeed Europe. As change was absolutely necessary to ensure a sustainable future, care was taken to deliver the message that heritage is not a barrier to betterment. While the most rigorous standards have been applied in the protection and conservation of Scheduled Ancient Monuments and Listed Buildings, less rigorous requirements have been prescribed in other situations. However all project work has been subject to very careful evaluation against the historic values and authenticity of the site.

It has also become increasingly obvious that the term 'cultural landscape' implies a close link between cultural and natural features, for example, the revegetation of the former mineral extraction areas is important in ensuring the structural integrity of the historic mineral workings. The ironworking and coal extraction that took place here was only possible because of the locality's geology, the presence of ironstone, coal, limestone, fireclay and the plentiful water supply. An appreciation of geology is therefore of fundamental importance to understanding the industrial past.

The contribution of voluntary organizations is very effective in the management of the World Heritage site. Blaenavon has benefited greatly from the various voluntary groups who have all assisted in protecting and promoting the values of the site.

Safeguarding a sustainable future

Since 2010, the Forgotten Landscapes Project has provided a holistic approach to heritage conservation management in the wider cultural landscape. The project, principally funded through the UK Heritage Lottery Fund, has been developed through extensive community consultation and adopted by the

Environmental improvements and special events have helped to increase the vibrancy of the town centre, offering a more sustainable future for local residents and traders.

Blaenavon World Heritage Site Partnership as it contributes to the ongoing protection of the outstanding universal value.

Implementation began with a specialist project team supported by Partnership members and an ever-growing army of volunteer rangers. Their work varies from heather moorland management and the recovery of the red grouse population, through access improvements, drystone wall repair and guided walks, as well as conservation and maintenance of key archaeological features within the World Heritage site. These activities are linked to lifelong learning courses, so that local people and visitors can continue to gain a better understanding as to why the site is of global importance.

Maintaining the involvement of the volunteer rangers in the longer term is critical to the sustainable future of the site. For this reason, a micro-hydro generation plant has been installed on the nearby River Afon Lwyd. The income earned from electricity sales will be used to resource this voluntary 'heritage workforce' and the ongoing conservation of the industrial landscape.

Work by the Forgotten Landscapes Project has also begun to prepare the case for the possible inclusion of a buffer zone to assist in the protection of the Blaenavon Industrial Landscape outstanding universal value and its setting. Although the case for a buffer zone was not made in the original nomination documents, the Management Plan review process has indicated the need for it and the benefits it would bring to the wider communities.

Positive benefits

There is no doubt that the investment undertaken in Blaenavon has achieved positive benefits for the community, the environment and the local economy since inscription on the World Heritage List. World Heritage status has provided a focus around which to rally local pride and sustainable regeneration and has been the catalyst for the many positive changes in Blaenavon over

299

Older residents reminisce at Blaenavon World Heritage Centre. The Centre engages with the local community, young and old, to celebrate the globally significant heritage of the area.

the last decade. Local people increasingly embrace their rich industrial heritage more readily; a clear demonstration of this was the community's recent decision to name their new school Blaenavon VC Heritage Primary School – a wonderful endorsement by the community of the Partnership's hard work.

Blaenavon continues to face many economic and social difficulties and the UK economic forecast for the coming years will no doubt present numerous challenges for the Partnership, which is committed and will continue to work towards a better future for this small, attractive Welsh town which played such a significant part in Britain's industrial development in the 18th and 19th centuries.

25

World Heritage in poverty alleviation: Serra da Capivara National Park, Brazil

ANNE-MARIE PESSIS, NIÉDE GUIDON AND GABRIELA MARTIN[1]

Treasures of the Piauí

In the north-east of Brazil connected with the Piauí and Bom Jesus do Gurgeia regions, there is a physiographical meeting point between the plateaus that make up a chain of *serras* (cliffs, ridges) and an ancient plain that is the peripheral depression of the middle São Francisco, the most important river in the region. The area is a major watershed, including the river valley system of Riacho Toca da Onca, Riacho Baixo da Lima, Riacho Bom Jesus and the Gruta do Pinga. The semi-arid landscape here is typical of the transition zone between the central and the Atlantic provinces. An abrupt *cuesta*, cliffs up to 270 m high, forms the border between two contrasting geological zones: a plain to the south-east and mountain massifs to the north-east. A network of narrow gorges and canyons formed due to erosion caused by rivers stretches for almost 180 km. On the sandstone walls on both sides of these gorges, rock shelters were formed, in which are preserved the traces of paintings and engravings that make up their great archaeological wealth.

The recognition of this unique landscape of treasures led to the establishment of Serra da Capivara National Park in 1979. The goal was to preserve

---

[1] The authors are researchers from the following universities: Anne-Marie Pessis, Fundação Museu do Homem Americano (FUMDHAM) and Universidade Federal de Pernambuco (UFPE, Brazil); Niéde Guidon, École des Hautes Études en Sciences Sociales (EHESS, Paris, France); Gabriela Martin, Fundação Museu do Homem Americano and Universidade Federal de Pernambuco (Spain).

Aerial view of museum and laboratories. The establishment of a non-profit foundation in 1986 and the construction of the Museum of the American Man Foundation, including research laboratories, had by 1990 consolidated conservation efforts in the park.

archaeological vestiges of what is believed to be the most ancient settlement in South America. It covers an area of about 130,000 ha in the south-east of the state of Piauí and includes the cities of São Raimundo Nonato, São João do Piauí, Canto do Buriti, Brejo do Piauí and Coronel José Dias. A 10 km long Permanent Preservation Area (APP: Área de Preservação Permanente) was created around the park, forming an additional buffer zone. The park, whose demarcation was completed in 1990, is dependent upon the Brazilian Environment and Renewable Natural Resources Institute (IBAMA). It was inscribed in 1991 on the World Heritage List for the magnitude and expressiveness of the geological and archaeological manifestations of the site, the exuberant landscape and the many species of plants and animals that have adapted to it. It was included on Brazil's Archaeological, Ethnographic and Scenic Inscription List in 1993.

The high density of prehistoric paintings and engravings include human figures, objects, animals and other figures that are not recognizable. There are numerous assemblages of figures representing scenes of ceremonial and everyday life. Until now 1,334 sites have been identified, 735 with paintings, 87 with engravings and 206 with paintings and engravings. The remaining sites are villages, camp-sites and lithic workshops and debris sites where the lithic industry shows a high technical level of knapping. There is also other very ancient evidence of ceramics and polished stones.

Research as cultural action

The National Park largely consists of dense thorny scrubland vegetation, known as *caatinga*, with a predominance of semi-arid vegetation dominated by succulents, drought-resistant deciduous thorny trees and shrubs, and other xerophytic vegetation. Relict isolated patches of forest cover survive in a few deep, narrow canyons. This vegetation, which includes palaeo-endemic relict

Canabrava child buried in an urn. The goal when setting up Serra da Capivara National Park in 1979 was to preserve archaeological vestiges of what is believed to be the most ancient settlement in South America.

genera and families representative of rainforest which were found in the area during the humid Ice Age of over 11,000 BP, is restricted to the canyons that retain moisture during the dry season. Serra de Capivara is recognized as one of the few protected areas within the *caatingas* biogeographic province which includes a vegetation type endemic to northeast Brazil. It contains unique species of animal and plant unknown elsewhere. Characteristic fauna is scarce in *caatinga* thorn scrubland, although recorded in the park are notable species including ocelot, bush dog, rocky cavy, red-legged seriema and a species of *Tropidurus* lizard.

Research shows that 70 million years ago the region was covered by the sea. Later, large animals – mastodons, super llamas and sloths up to 8 m tall – dominated the landscape. The region was continuously occupied by hunters/gatherers and agriculturists/ceramists until the arrival of the colonizers in the 16th century. The study of the paintings inside the shelters allows for the identification of traditions, ceremonies, myths, rituals and scenes of daily life dating back 12,000 years. The mission to Serra da Capivara National Park and more precisely to the Boqueirão da Pedra Furada archaeological site, discovered 50,000-year-old vestiges of *Homo sapiens*. The discovery raised an international controversy as it refuted the theory accepted for over half a century that man had reached the American continent only 12,000 to 15,000 years ago.

Since 1973, an archaeological research programme has been under way in São Raimundo Nonato, in the south-east of Piauí. It became interdisciplinary in 1978 with the formation of the French-Brazilian Mission of Piauí, archaeologists, ecologists, zoologists, botanists and pedologists among other professionals. The research focus is 'The interface between mankind and the environment – from prehistory to the present day in the Serra da Capivara National Park'. The researchers have been carrying out surveys and studies of the archaeological sites, taking into account the geological features, the fauna, the flora and human occupation from the outset to the present day.

303

The diagnosis of pathologies of rock art sites through geochemical and microbiological procedures is obtained by the results of regular monitoring conducted by the Laboratory of Historic Monuments and Protected Areas of the Direction de l'Architecture du Patrimoine de France and the Laboratory of Archaeological and Heritage Metrology of the Universidade Federal de Pernambuco. Effective monitoring of the rock art has enabled a rapid response with possible preventive and remedial measures.

The establishment of a non-profit foundation in 1986 and the construction of the Museum of the American Man Foundation (FUMDHAM: Fundação Museu do Homem Americano) including research laboratories had by 1990 consolidated conservation efforts in the park. The museum is responsible for protecting, preserving and disseminating information about the material found in the course of the archaeological, anthropological and ecological research pursued in the region. Items in the museum include the jaw of a sabre-tooth tiger, a fragment of fossilized skull, an 8,960-year-old ceramic piece and a 9,200-year-old polished stone axe.

Infrastructure for community engagement

The development of the museum was essential to increase visitation. Apart from the permanent exhibition based on research, a series of temporary exhibitions are also produced. The museum project enabled the training of laboratory technicians to support research in the region. It was also an incentive to attract researchers from Brazil and abroad. The construction work created jobs for a population that had no income and allowed a participatory process of integration of the local population into the conservation of heritage resources. A network of roads within the park was built, allowing access for researchers, guards, and the fledgling number of visitors.

Technical cooperation from Italy for an educational development project led to the creation of a number of community support centres around the perimeter of the park. In the rural areas where there were neither schools nor any other public body, primary schools were built along with health units, which reduced the high rate of infant mortality. The involvement of teachers educated at universities of São Paulo ensured a good education, and improved teacher capacity at schools in urban areas of the region. Researchers at the Oswaldo Cruz Foundation in Rio de Janeiro ensured that there were adequately trained personnel working in the health units. These education and health activities began to lay the foundation for the promotion and the involvement of the local society in conservation.

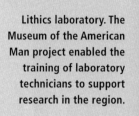

Lithics laboratory. The Museum of the American Man project enabled the training of laboratory technicians to support research in the region.

305

Adult education was promoted among the rural communities surrounding the National Park in bee-keeping and handcrafting pottery. The region is rich in diverse clay types. It is also abundant with honeybee populations that are attracted by bee houses. Honey-collecting and filtering is performed in a specialized local laboratory. These activities increased family income. By the end of the five-year Italian cooperation there was increased awareness on the part of the communities bordering the National Park about the prospects of diversifying livelihood, the importance of archaeological heritage and the value of social partnerships for addressing Millennium Development Goals.

At the FUMDHAM headquarters, conferences and seminars are held annually with the participation of researchers and graduate students. There are numerous doctoral students who remain in the region to carry out fieldwork. These ongoing activities for more than two decades have spurred the Ministry of Education to create a campus of a federal university, establishing the first degree in archaeology in Brazil. Today, the first graduate classes are a reality. Many of them are from the region and wish to stay in the locality and participate in the research activities. The teacher/student dynamics, as well as the studies carried out at the FUMDHAM headquarters, have become an engine for development in the region, having its social landscape radically changed. To support the development of education in archaeology, a natural sciences undergraduate degree was created, thereby complementing the training of researchers who will be able to continue the work begun in the 1970s.

Museum of the American Man. The development of the museum was essential to increase visitation. Apart from the permanent exhibition based on research, a series of temporary exhibitions are also produced.

Managing at the margins

The management of Serra da Capivara National Park, by law, must be exercised by the Ministry of the Environment. FUMDHAM have a partnership agreement with the Ministry, being a Public Interest Civil Organization, governed by specific legal instruments which depend on the Ministry of Justice. FUMDHAM with the mandate of the federal authorities is responsible for entry control to the park and its maintenance. Sponsorship provides for the operational functionality of the park, which is regularly negotiated. Inside the park the conservation status of the roads and sites is of excellent quality. There are twenty-four access gatehouses with services for visitors, who must be accompanied by a guide from the National Park. These guides are trained by FUMDHAM and are regularly rostered on schedules.

Studies conducted on the self-sustainability of FUMDHAM and of Serra da Capivara National Park underline the importance of developing heritage tourism activities. In the sector with the highest concentration of sites containing rock paintings, five circuits of public visits have been established: Baixão das Mulheres, Desfiladeiro da Capivara, Baixão da Pedra Furada, Sítio do Meio and Baixão do Perna.

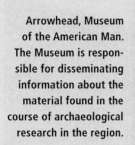

Arrowhead, Museum of the American Man. The Museum is responsible for disseminating information about the material found in the course of archaeological research in the region.

An acceptable road infrastructure outside the park, a good airport and local hotels are necessary to develop sustainable tourism, and an adequate transport system is needed for building hotels around the park. Currently the shortest and most direct route to reach a regular airport is through 300 km of totally impassable roads that require maintenance resulting from the negligence of the road authorities. The Serra da Capivara Airport was completed in 2009 with funds from the federal government and built by the state government of Piauí. The airstrip is excellent, and the airport will become active as soon as the passenger terminal building is finished. This infrastructure will be critical for any project of self-sustainability. Until now the lack of access by road or air has discouraged any kind of investment, but it is envisaged that this situation will change soon.

Despite the problems encountered and surveyed, everything that has been accomplished during these decades in Serra da Capivara, has shown that investing in human resources and scientific research have multiplier effects in economic and social development of economically depressed areas. All that remains is to solve in a more efficient manner the problems created by distance, demand for infrastructure and grassroots poverty.

Many initiatives have been taken to integrate the local population into the archaeological heritage conservation project. The park's Management Plan prepared by FUMDHAM researchers considered the present population as one of the elements of ecosystems that should be integrated into all aspects of interventions and developments. Hence measures to improve education, health and employment were adopted, which steadily changed the living conditions of extreme poverty that existed when research began in the region. The local community has benefited from the establishment of a federal university

Serra da Capivara National Park. The main body of the park is the Serra do Congo massif and the central Chapada da Capivara in the state of Piauí.

The local community has benefited from the establishment of a federal university that educates archaeologists and professionals in several natural science subjects. These young people will continue the work of FUMDHAM. The park provides work in a region where droughts prevail and where agriculture is not profitable.

that educates archaeologists and professionals in several natural science subjects. These young people will continue the work of FUMDHAM. The park provides work in a region where droughts prevail and where agriculture is not profitable.

A budget of 2 million euros is sought from the federal government for the sustainable development of the National Park. This will eventuate once the airport is opened and with the establishment of the hospitality infrastructure. It is envisaged that visitors will be able to access the region, contributing to the growth of tourism. Donations from cooperation with Petrobrás have been extremely helpful to date.

Transformations through outstanding universal value engagement

The development of FUMDHAM in Serra da Capivara initially faced the challenge of lack of awareness among the local communities about the outstanding universal value of the World Heritage site and its relevance for local community development. The development and implementation of educational programmes have been critical for promoting awareness of the symbolic value of the site, which is significantly different from a monetary perspective. Others programmes such as training for craftwork production (pottery) and the

Construction work created jobs for a local population that had no income and allowed their integration into the conservation of heritage resources.

establishment of a honey cooperative have helped to build the trust and commitment of the local communities.

Community engagement in the protection of heritage sites results is complex. Bringing together host communities and their heritage sites demands appropriate capacity building for ensuring economic or financial activities and community benefits. It includes developing tourism infrastructure, up-skilling for integrated local area planning and investment in small-scale local enterprises that benefit both the heritage site conservation and local populations.

It is demonstrated that to link the potential benefits for local people from living next to historical, artistic and archaeological attractions and the sustainable conservation of heritage sites involves the participation of researchers, local authorities, and the primary stakeholder people who live around them. Issues of local level of involvement, taking ownership for the conservation of the site as local custodians, the mix of local and outside control of businesses, and ideas and concerns about managing the site are important factors.

309

The challenges are related to improving schooling, a decision-making process based on relevant criteria and a timeline and the adoption of more autonomous approaches to financial improvement. The experience of FUMDHAM has proved the need for inclusive participatory processes and that this would last longer once the impacts and benefits are experienced.

Research and development in Serra da Capivara National Park is focused on capturing high-resolution digital photographs, linked to the creation of hybrid digital models of sites with rock paintings. When the procedure is complete it will form a collection of colorimetric, microscopic fluorescence and X-ray data of the prehistoric rock art. The processed three-dimensional digital models data, measurements performed on created models and chronological, colorimetric and microscopic information are integrated into a database. The goal is to be able to perform this type of documentation in all sites with rock paintings and engravings.

Research and development in Serra da Capivara National Park is focused on capturing high-resolution digital photographs, linked to the creation of hybrid digital models of sites with rock paintings.

Outstanding universal value – catalyst in poverty alleviation

The technical and financial support of the Interamerican Development Bank was critical to make operational the steps for the establishment of the National Park as an immediate strategy to slow the process of degradation

Excavations. Over 300 archaeological sites have been found within the park, the majority consisting of rock and wall paintings dating from 50,000–30,000 years ago.

310

Prior to the establishment of the National Park, the region was poor, the infant mortality rate was high and there were no basic public utilities such as water, electricity and communication. The change has been astonishing with all of the public utilities and services now available.

of the World Heritage site. Experience at Serra da Capivara has shown that educating younger people from the local population, giving technical training to adults and providing centres for public health are essential to integrate the populace into archaeological heritage conservation and research activities. The degradation of the archaeological heritage caused by natural agents has also been reduced as a result of the implementation of interventions for diagnosis and regular conservation of rock art sites. Intergenerational integration of the region's population through both formal and informal education has become a reality and this is relied upon as an alliance in the preservation of archaeological and natural heritage.

These outcomes are the result of more than two decades of combined efforts between researchers from several institutions who have accepted the challenge. Turning research into a tool for socio-economic development in a deprived region is a mechanism for transformation. Sustainability now depends on finishing current projects: the airport and the diversification of the infrastructure for developing the hospitality industry and responsible tourism. To guarantee this transition a fixed federal budget for the conservation of Serra da Capivara National Park is required.

Prior to the establishment of the National Park, the region was poor, the infant mortality rate was high and there were no basic public utilities such as

Many of the numerous rock shelters in the Serra da Capivara National Park are decorated with cave paintings, some more than 25,000 years old.

311

water, electricity and communication. The change has been astonishing with all of the public utilities and services now available. There is a state university, a federal university, a federal technical training centre, a hospital and three banks in the region. FUMDHAM has been the catalyst for development and it is the largest employer. Priority is given to the involvement of women in the sustainable development of the project. FUMDHAM prefers employing women for park vigilance, to control entrances open to visitors and also for outsourced services such as souvenirs shops and snack counters. Women are also employed for technical and professional activities in the laboratories and research centres. They also constitute the majority of FUMDHAM employees, of which there are currently 113 – 93 women and 20 men.

Some of the next steps are to be programmes that enhance 360-degree knowledge about the heritage site, training programmes on entrepreneurship, marketing, access to capital for small business, and education of both planners and local people. Such activities are part of strategies at local, national and international levels, and they will be compatible with the principles of the 1992 and 2012 Rio Earth Summits.

26

Angkor Archaeological Park and communities: Angkor, Cambodia

KHUN-NEAY KHUON[1]

Sustainable development in a historic area

Angkor Archaeological Park, covering an area of some 400 km², comprises 112 villages spread across the boundaries of the World Heritage site plus Siem Reap with its substantial tourist infrastructure and provincial administration. The Royal Khmer Government has made a firm commitment to secure the livelihood of the local population as an integral part of the park and will be actively engaged in park management and the benefits from community-based development. In 2008 restructuring of the APSARA Authority[2] led to the formation of the Department of Land and Habitat Management in Angkor Park, which is responsible for analysis, evaluation, monitoring and actions closely related to the sustainable development of the population in this historic area. This case study focuses on achievements to date.

The total population of the park was about 120,000 in 2010. Each inhabitant has on average 1 ha of agricultural land. The annual rice production rarely reaches two crops. The shortage of food is supplemented, especially around the months of September and October, with rice purchased elsewhere. The income of the local communities has been broken down as follows: hand collecting firewood (27 per cent), rice cultivation (20 per cent), unskilled jobs (17 per cent) and several other minor activities (36 per cent). Literacy in the park stands at 32 per cent with a 2:1 ratio of men to women. The majority of families (about

---

[1] Architect and planner, Deputy Director-General, APSARA Authority, Cambodia.
[2] Authority for the Protection and Management of Angkor and the Region of Siem Reap.

Restoration works of the Southern Bayon Library.

60 per cent) are poor. According to a study dated 2007, their average monthly income per capita was US$24–30. Things began to change for the better through sustainable development defined by the continuity and participation of local communities in the conservation of the World Heritage communities.

Appropriate governance

APSARA manages the park in consultation with the International Coordinating Committee (ICC-Angkor) created in 1993 and co-chaired by France and Japan and for which UNESCO provided the Secretariat. ICC-Angkor is an international coordination mechanism that ensures the sharing of information between APSARA and the international teams working at Angkor, sets technical standards where necessary and encourages international cooperation in the fields of conservation and sustainable development.

The first Angkor Management Plan was developed in 2007 with assistance from New Zealand. A second tool, Angkor Heritage Management Framework, was set up by the end of 2010 in collaboration with UNESCO and thanks to the financial support of Australia. Both tools will be integrated to become the final Angkor Management Plan.

The Authority has adopted a consultation mechanism for each major development project through a steering committee of specialists, representatives of local authorities and private sector, representatives of the local population living in the park, and the Buddhist clergy.

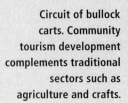

Circuit of bullock carts. Community tourism development complements traditional sectors such as agriculture and crafts.

The 1996 Law on Protection of Cultural Heritage and the 2001 Land Law inform the park management. Government Decision No. 70 SSR 16 September 2004, on land use in Angkor Park, while confirming that provision, states that long-time residents of the protected Zones 1 and 2 in Angkor Park can continue to live there without being evacuated. They can renovate and even replace their old houses with new ones and bequeath them to their descendants. But they cannot add more homes or sell the land to outsiders. APSARA has legal backing in the Royal Government Circular No. 01/SR of 6 May 2004 for peaceful living in Angkor Archaeological Park. This was further reinforced a year later with the Order of the Royal Government 02/BB. The enforcement of these provisions was consolidated in 2008 through the Department of Public Order and Cooperation, equipped with the necessary staff and equipment. This new department functions in close cooperation with the local authorities.

One of the important projects now being implemented is the Angkor Participatory Natural Resource Management & Livelihood (APNRM&L). The sponsoring partner, New Zealand, is currently training fifteen liaison officers. They will be allocated to each district as vital conduits for a direct working relationship between the village communities and APSARA. The villagers living in the park are assured of continuity of residence by the Authority through

Forces of the Department of Public Order on patrol. APSARA has legal backing from the government and this was reinforced with the Department of Public Order and Cooperation, equipped with the necessary staff and equipment.

the establishment of land-use plans with their participation and through land registration. Certificates of occupancy of land are issued to residents of long standing. A specialized unit, equipped with modern facilities (aerial photographs, GIS maps, electronic topography devices, registration system networking, etc.) has been actively facilitating this process.

Run Ta-Ek ecovillage

The minimization of impacts on the outstanding universal value of the World Heritage property is assured through restricting the population to current levels by encouraging younger people and new families to move outside the protected areas on a voluntary basis. The Authority has acquired 1,012 ha in the town of Run Ta-Ek to the east of the park, about a half-hour drive from Siem Reap. The new settlement will offer future residents a sound infrastructure of roads, agricultural land, an irrigation system, micro-credit facilities, schools, a vocational training centre and a Buddhist monastery, as well as solar energy, windmills to pump water from lakes, organic agriculture and ecotourism networks. As it is outside the protection zone of Angkor Park, homestay for visitors in people's houses and other recreational facilities are strongly encouraged to enable villagers to obtain additional income for their families.

The new settlement will offer future residents a sound infrastructure of roads, agricultural land, an irrigation system, micro-credit facilities, schools, a vocational training centre and a Buddhist monastery, as well as solar energy, windmills to pump water from lakes, organic agriculture and ecotourism networks.

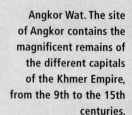

Angkor Wat. The site of Angkor contains the magnificent remains of the different capitals of the Khmer Empire, from the 9th to the 15th centuries.

The ecovillage is organized around natural lakes in five housing units, which accommodate 850 families or about 5,000 people. Each family receives a hectare of land for the construction of their house and for agricultural activities. The first unit, called Chea Lea, is planned for 200 families. For the first 100 families, the government grants substantial aid: $10\,m^3$ of timber, concrete pillars, sheets for the roof and a toilet bowl and equipment for the septic tank. Currently, ninety households have been set up. A school with five classrooms has been constructed and a large agricultural station is in operation and recruiting new residents, who receive a bursary and technical training in organic farming which they can apply in their own fields.

Community participation and benefits

The APNRM&L demonstration project enables sustainable development through the local population in Angkor Park to ensure equitable income generation through development of local resources, diversification of economic activities and training for cultural tourism, which aims to facilitate direct tourism revenues to the poor without intermediaries.

Participatory democracy is ensured through direct consultative sessions with the community with the following objectives:

- Building trust between APSARA and the local population, creating a climate conducive to a true partnership.
- Increasing opportunities to improve the living conditions of villagers from sustainable use of natural resources and tourism.
- Strengthening the role and capacity of villagers to take responsibility for their futures and actively participate in sustainable development of the park.
- Establishing a Village Development Committee by the villagers through direct elections, with subcommittees for specific programmes in agriculture, livestock, forest management, crafts, micro-credit, drinking water, renewable energy and so on.
- Strengthening the capacity of APSARA to enable it to fulfill its mandates and achieve the above objectives.

The APNRM&L demonstration project implemented in the second half of 2010 has been evaluated by the New Zealand Aid Programme (NZAID) with successful outcomes against the objectives. The villagers have taken ownership of their participation in the conservation of the park. As a result NZAID has further resourced development from January 2011 and extended it to eight villages within the park – a good case study on the multiplier effect of donor support in sustainable development.

317

The APSARA Authority is also working with the National Federation of UNESCO Associations in Japan to establish community learning centres. The primary goal is to develop and strengthen the capacity of village communities through non-formal education and vocational training.

Community tourism Community tourism development complements traditional sectors such as agriculture and crafts. A circuit of bullock carts and a craft shop are directly managed. APSARA has also started the construction of a wooden bridge, two viewing platforms and the installation of signs. Bicycle tours are planned between villages and Scout camps are being established. As overnight stays are not allowed in the park, community-driven tourism focuses on rustic and culinary experiences.

The management of tourism activities is entirely driven by community-elected committees that take full responsibility. Income is divided into four components for each activity: direct fees for owner-drivers of carts, or artisans; equipment maintenance; fees of the respective committees; and fund for village development. This fund is intended to work for various common benefits of the whole village, such as improving irrigation channels and communications.

The goal is to integrate intangible heritage in the interpretation and management of the tangible heritage. Guidelines for safeguarding intangible heritage in Angkor Park have been drafted through the participation of local communities.

Living heritage APSARA has commissioned research and development projects on intangible heritage, especially village life and monastic traditions and lifestyles. The goal is to integrate intangible heritage in the interpretation and management of the tangible heritage. Guidelines for safeguarding intangible heritage in Angkor Park have been drafted through the participation of local communities.

Sustainable development

The ICC-Angkor established a group of ad hoc experts in 2006 to contribute to reflection and implementation of sustainable development projects within and around the Angkor site, which mainly involved local communities.

Agriculture APSARA permits are required for agricultural development as a key economic driver for the local communities. Organic farming and improved and appropriate technologies are facilitated to increase farm profitability with minimal impacts on the environment. Composting methods, botanical pesticides and the KEM (Khmer Effective Micro-organisms) from local raw materials are being developed. Agricultural experiments on pilot plots have demonstrated the effectiveness of KEM on vegetables and rice. Excellent results were also obtained in pilots on poultry and fish farming. In order to fully implement these innovative practices, peasant associations are

Local children fishing in pond, Angkor Wat.

encouraged and a team of seventeen development officers has been appointed with responsibility for agricultural development.

Water management Angkor Park is located between Mount Kulen and Tonle Sap Lake. Floods can reach the city at any time during heavy storms and rain. Ancestral practices to avoid these disasters included a system of flood spillways and dykes. All Khmer temples in the flat areas were surrounded by a moat, for collecting water flow from the temple and to recharge the sand layer under the temple, making it more robust to support the structures.

In order to ensure a water supply to the entire region and the conservation of Angkor Park, an integrated management system of water resources has been adopted. Old systems are repaired and new ones built for irrigation and managing floods. The collaboration of local people has become vital, with a trickle-down of benefits. Water from the temple moats is also diverted to neighbouring village communities for the irrigation of their rice fields.

319

Forest management The area of forest cover amounts to 4,574 ha, representing 11.44 per cent of the total area of Zones 1 and 2 of the protected area in Angkor Park. Trees have an intimate relationship with the cultural landscape and security of temples. Their shade and their roots provide strength to the stones and loss of the forest in the Angkor area would inevitably reduce the life of the temples. Since 2004 APSARA has invested considerable resources in forestry through:

- creating nurseries and distribution of seedlings to schools and local communities;
- studying the trees' biological characteristics for the treatment of disease and conservation of the historic landscape of the Angkor area;
- researching cleared land for organic afforestation;
- taking preventive measures against forest fires and promoting environmental protection to improve cultural landscapes;
- minimizing forest crime by recruiting 130 forest conservation agents to ensure permanent custody; and
- planning gardens of spices and medicinal plants to safeguard old traditions.

At the request of the villagers, community forestry is promoted and directly managed by them for their regular needs, with technical assistance from APSARA. Two villages have had community forests of 30 ha established since 2009.

Awareness programme A mobile team from the APSARA Department of Communication visits all the villages to give information to residents, school-children, students and religious establishments. Specific sessions are organized to raise awareness of the monks, police and APSARA staff operating in the field.

The Authority publishes a weekly magazine that serves to promote conservation values among communities and local authorities. A weekly programme of half an hour is broadcast through the local radio station in Siem Reap. APSARA is also associated with national TV programmes on a series of special topics relating to the management of Angkor Park and particular sites. These promotional activities started in 2004 and continue today.

Heritage interpretation

APSARA has built an interpretation centre in the village of Rohal along the access to the main temples. It is an important place for visitors to take a break and learn about communities in Angkor Park. Interpretive displays include a Khmer-style house along with models and pictures of different types of traditional houses, built around a vegetable garden, as seen in Khmer villages, with the different types of vegetables grown. The importance of indigenous plant foods is highlighted as the local restaurants and hotels use mainly imported

Bayon temple, 13th century, Angkor Thom. The influence of Khmer art played a fundamental role in the distinctive evolution of South-East Asia. The result was a new artistic horizon in Oriental art and architecture.

foreign vegetables. Local flowers and fruit trees are also presented. The centre is surrounded by hedges with shrubs that provide a variety of leaves, flowers and edible fruit for the preparation of traditional dishes. The villagers can also prune these shrubs for firewood. The centre thus plays a very important role in promoting sustainable development.

In addition, APSARA urges organizations working on the Angkor site to establish centres of interpretation specific to the temples of their project concern. Thus, Switzerland has an interpretation centre in the court of Banteay Srey, the Japan-APSARA Safeguarding Angkor (JASA) Centre has Bayon, there is the University of Sophia Center at Prasat Banteay Kdei, the World Monuments Fund has a visitor centre at Prasat Preah Khan, India's centre is at Prasat Ta Prohm, China's at Prasat Chau Say Tevoda, and the German and Angkor Conservation Project has its centre at Angkor Wat Prasat.

The Preah Norodom Sihanouk Angkor Museum opened in November 2007. It specializes in Buddhist art and exhibits the 274 pieces found by Sophia University in its conservation work at Banteay Kdei. Another museum, dedicated to ceramics, was built in December 2009 in collaboration with Japan, in the village of Tani. It profiles five groups of ceramic kilns dating from Angkor historical times. This new museum is intended as an archaeological park garden where a visit can be combined with recreational activities including walks, picnics and golf for children. The third museum, built in collaboration with India in 2012, is dedicated to Asian silk.

321

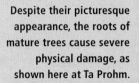
Despite their picturesque appearance, the roots of mature trees cause severe physical damage, as shown here at Ta Prohm.

Buddhist monks, Angkor Wat. One of the immeasurable impacts of the Angkor Park development is on the mental health of a devastated population, which has been able to regroup and believe in their sense of self as proud people.

Youth programmes

A Heritage Education programme started in 2009 with the preparation of a series of teaching materials in collaboration with the National Federation of UNESCO Associations in Japan. A project with the Siem Reap Department of Education is under way for implementation in schools. APSARA is responsible for the overall concept, development tools, training of trainers and coordination.

Colouring Book No. 1, Heritage Series, has been published for 3rd- and 4th-year students (aged 9–10) in 2009. In Siem Reap province, students from all twelve districts participated in colouring games. In addition, hundreds of copies were sent to departments of education in all Cambodian provinces. Encouraged by the school in Siem Reap, the three partners continued their collaboration during 2010, for sensitizing children in 5th and 6th years (aged 10–11). To this end, it is not only an exercise in colouring, but also and especially to get young people to rally behind the conservation of the Angkor World Heritage site. The second phase is at an advanced level. It includes a teacher resource book for preparing the pupils before, during and after a visit to the monuments. The new campaign covers twenty-five schools in the twelve districts of Siem Reap province, young Scouts and top students from other provinces. More than 1,000 copies have been sent to provincial education departments across the country. The programme is ongoing in 2012.

The success of the Heritage Education programme in schools prompted the Siem Reap Department of Education to formally incorporate World Heritage education in the school curriculum. Moreover, APSARA has provided assistance to non-governmental organizations such as Private Initiative Support for

View of Run Ta-Ek ecovillage. The ecovillage is organized around natural lakes in five housing units, which accommodate 850 families or about 5,000 people.

Aid to Reconstruction (SIPAR), specializing in school libraries and promoting reading through a new series of small books entitled *Exploring Angkor*. The first of the series was published in December 2011 in three languages, Khmer, English and French, thanks to the support of UNESCO.

323

Programme for local communities

The objective of the APNRM&L project in cooperation with New Zealand is to get the villagers to participate in the conservation and sustainable development of the park, by focusing on activities that can provide direct benefits to the public and improve living conditions. The heritage education for the villagers, which was in place by mid 2011, includes several steps:

- Focus on tools promoted with the participation of the key people of the targeted villages. The topics on development cover systems of knowledge based on the culture of Cambodia and the local intangible heritage. They converge on the identification of tasks that the villagers have to do individually and collectively to improve their conditions of life, and at the same time, protect the heritage.
- Training of trainers in APSARA among the staff concerned.
- Customized and localized training of villagers through the APSARA trainers.

The 2012 training programme is under way in the villages targeted by the APSARA–New Zealand project. Buddhist clergy are also actively engaged in the preservation of Angkor heritage and especially in the education of villagers. A two-day seminar was organized in October 2005 for the leaders of all the monasteries located in Angkor Park. His Holiness the Supreme Chief of Buddhism in Cambodia and the high dignitaries representing the Royal Government were present. APSARA conducted the second phase of training for selected monks in March 2012.

Regrouping of a proud people

The mutual understanding and trust between APSARA, park management and residents builds a new foundation for long-term collaboration.

The management of a large heritage park such as Angkor is a colossal challenge, and the location of the park just north of the city of Siem Reap border called for additional management interventions. But Cambodia has been able to take on the challenge despite a quarter century of war and the consequent physical, economic and human destruction. One of the immeasurable impacts of the Angkor Park development is on the mental health of a devastated population, who have been able to regroup and believe in their sense of self as proud people. The solidarity of the international community in setting up APSARA in 1995, with a few professionals, has gradually provided material, financial and human resources for appropriate sustainable development.

Alongside conservation activities, APSARA has attached great importance to the village communities in Angkor Park, especially since the restructuring of its operations in 2004. This new policy to integrate local communities in all stages of the management of the park has provided tangible results. What has become critical is the realization and ownership by the villagers of economic outcomes that they can obtain from the sustainable use of natural resources and their participation in the protection of the park. In the long term, this factor will ensure the benefits and successes of any other efforts and innovations introduced, developed and evaluated with the close collaboration of local, provincial, national and international stakeholders.

Pathways to sustainable development

KISHORE RAO[1]

Nearly half a century ago, the World Heritage Convention was conceived as a mechanism for international cooperation to safeguard sites of universal significance and it has since become an all-inclusive instrument of international heritage law, to bring diverse cultures and their heritage together. Today, with 190 States Parties, the benefits of this vision are increasingly perceived and shared beyond the borders of the World Heritage sites, underscoring the importance of the concept of outstanding universal value as a vector of sustainable development, particularly local community development. The holistic conservation ethic is succinctly epitomized in the overall theme of the 40th anniversary of the Convention, celebrated in 2012 as *World Heritage and Sustainable Development – the Role of Local Communities*. Shared understanding of outstanding universal value among culturally and linguistically diverse communities around the world has become a cultural right, true to the spirit of the Convention that is at once shared by all stakeholders at local, regional, national and global levels. Most significantly, it has become a crucial contributor to sustainable development.

This spirit and commitment was emphasized by the Director-General of UNESCO, Irina Bokova, launching the 40th anniversary celebrations, when she called for a year of renewal for World Heritage. 'Heritage stands at the crossroads of climate change, social transformations and processes of reconciliation between peoples. Heritage carries high stakes – for the identity and belonging of peoples, for the sustainable economic and social development of communities.'[2] She argued that 'heritage does not represent luxury; it is a capital investment in the future. It is the sound foundation without which nothing lasting can be

---

[1] Director, UNESCO World Heritage Centre.
[2] Opening address to the General Assembly of the States Parties at the World Heritage Convention, 7 November 2011.

built. Disregarding heritage, severing our root, will inevitably clip our wings.' She has consistently advocated for a paradigm shift to further sustainable development, 'a new approach to research that is interdisciplinary, solutions oriented and policy relevant, with a stronger social science component.'[3]

As Ban Ki-moon, Secretary-General of the United Nations, put it: 'Global economic growth per capita has combined with a world population (passing 7 billion last year) to put unprecedented stress on fragile ecosystems. We recognize that we cannot continue to burn and consume our way to prosperity. Yet we have not embraced the obvious solution – the only possible solution, now as it was twenty years ago: sustainable development.'[4]

Redefining human development

The strategies that have been followed so far, indeed, seem to have exposed their limits. Clearly, there is a need for a new approach to development and sustainability, one that would work. One way of addressing this challenge is to redefine what we mean by human development. In this regard, there is widespread agreement that 'progress needs to be defined and measured in a way which accounts for the broader picture of human development and its context', which would emphasize 'equity, dignity, happiness, sustainability' (Helen Clark, United Nations Development Programme administrator).[5]

In its report *Resilient People, Resilient Planet*,[6] the UN Secretary-General's High-Level Global Sustainability Panel concluded that 'the international community should measure development beyond GDP and develop a new sustainable development index or set of indicators'. These views are also reflected in the Organisation for Economic Co-operation and Development's Better Life Index initiative[7] and the Stiglitz-Sen-Fitoussi Commission[8] and numerous other similar initiatives, which called for a broad range of social indicators to complement GDP figures. The recently held United Nations Conference on Sustainable Development, also known as Rio+20, endorsed this idea by recognizing '... the need for broader measures of progress to complement GDP in order to better inform policy decisions ...'.

---

[3] Address to *Preparing the way to sustainable development after Rio+20: Forum on Science Technology and Innovation for Sustainable Development*, 27 June 2012.

[4] The Future We Want, *International Herald Tribune*, 24 May 2012.

[5] High-level forum at the UN Conference on Sustainable Development, Rio de Janeiro, Brazil, 20 June 2012.

[6] http://www.un.org/gsp/report/

[7] http://www.oecdbetterlifeindex.org/

[8] Commission on the Measurement of Economic Performance and Social Progress: http://www.stiglitz-sen-fitoussi.fr/en/index.htm

Recognizing the role of heritage in sustainable development

Cultural and natural heritage is certainly a major consideration when it comes to defining the constitutive elements of well-being, dignity and sustainable human development in general. UNESCO has been promoting, at least since the 1980s, the role of culture – and heritage – as a driver and enabler of development.[9] Recently, these efforts have resulted in two landmark resolutions adopted by the General Assembly of the United Nations, N. 65/166 and N. 66/208, which emphasize the crucial importance of culture as 'an essential component of human development, a source of identity, innovation and creativity for the individual and the community'.[10]

Rio+20 has reiterated this concept by recognizing that 'all cultures and civilizations can contribute to sustainable development' (Para. 41 of the outcome document, *The Future We Want*)[11] and that 'many people, especially the poor, depend directly on ecosystems for their livelihoods, their economic, social and physical well-being, and their cultural heritage' (Para. 30). The Rio Conference has also stressed the 'intrinsic value of biological diversity, as well as its ecological, genetic, social, economic, scientific, educational, cultural, recreational and aesthetic values' (Para. 197).

327

Linking cultural and natural heritage

There is increasing recognition that, at a fundamental level, biological and cultural diversities are closely interdependent. They have developed over time through mutual adaptation between humans and their environment, and therefore, rather than existing in separate and parallel realms, they interact with and affect one another in complex ways in a sort of co-evolutionary process. For this reason, traditional and indigenous practices for the stewardship and use of environmental resources, including building techniques, are in general green by design. They embody an intrinsically more sustainable pattern of land use, production and consumption, contributing also to food security and water conservation and access, based on knowledge and practices developed over centuries of adaptation. This suggests that any local policy

[9] http://www.unesco.org/new/en/culture/themes/culture-and-development/the-future-we-want-the-role-of-culture/

[10] http://daccess-dds-ny.un.org/doc/UNDOC/GEN/N10/522/50/PDF/N1052250.pdf?OpenElement;

[11] http://www.uncsd2012.org/thefuturewewant.html

aiming to protect the natural environment and achieving sustainable development will also necessarily have to take into consideration, and act upon, the culture of the communities concerned.

Integrating heritage in the post-2015 development agenda

In 2015, the international community will review the progress made in the achievement of the Millennium Development Goals (adopted in 2000), and define a new set of Sustainable Development Goals (SDGs), which will set priorities for the post-2015 development agenda. In this context, it is crucial to ensure that the conservation and wise use of heritage, both natural and cultural, is taken into account and fully integrated in future sustainable development policies and programmes. The coming years will provide in this regard a unique window of opportunity to make and justify such a case. However, this will require a strong and consolidated set of experience and evidence-based arguments.

The contribution of the World Heritage programme

World Heritage, a flagship programme of UNESCO over the past forty years, is well placed, thanks to its visibility and popularity, to illustrate the extraordinary contribution that the conservation of outstanding cultural and natural properties has made and could make to sustainable development. Almost a thousand sites, including vast marine reserves, extensive tropical forests, historic cities and living cultural landscapes in all regions of the world, have been identified, protected, managed and monitored over a number of years, resulting in a wealth of experience and important data on the relationship between heritage and sustainable development.

The importance of this link has been recognized on a number of occasions by the states that have ratified the World Heritage Convention. Most recently, the 18th General Assembly of the States Parties (Paris, 2011) has adopted a Strategic Action Plan for the Implementation of the Convention, 2012–2022, which calls for the 1972 Convention to 'contribute to the sustainable development of the world's communities and cultures' by ensuring that 'heritage protection and conservation considers present and future environmental, societal and economic needs'. As noted above, the importance attached to this issue is also reflected in the official theme chosen for the celebrations of the Convention's 40th anniversary in 2012.

Indeed, as an attribute of natural and cultural diversity, World Heritage plays a fundamental role in fostering sustainable development and as a source of well-being. Through a variety of goods and services and as a storehouse of knowledge, a well-protected World Heritage property contributes directly to providing basic goods, security and health, providing access to clean air, water, food and other key resources as well as attracting investments and ensuring green, locally based, stable and decent livelihoods, only some of which may be related to tourism.

Most activities associated with the stewardship of cultural and natural heritage, developed over centuries if not millennia of slow adaptation, have indeed a much lower impact on the environment compared with other sectors, while generating sustainable local employment opportunities, including the fostering of creative industries based on local arts, crafts and other products. This is true of natural protected areas rich in biodiversity, of course, but also of cultural landscapes and historic cities. A well-maintained heritage is also very important in addressing risks related to natural and human-made disasters. Experience has shown how the degradation of natural resources, neglected rural areas, urban sprawl and inadequately engineered new constructions increase the vulnerability of communities to disaster risks, especially in poorer countries. On the other hand, a well-conserved natural and historic environment considerably reduces underlying disaster risk factors, strengthens the resilience of communities and saves lives.

World Heritage, finally, is essential to human spiritual well-being for its powerful symbolic and aesthetic dimensions. Conservation of the diversity of the cultural and natural heritage, fair access to it and the equitable sharing of the benefits deriving from its use, enhance the feeling of place and belonging, mutual respect for others and a sense of purpose and ability to provide for succeeding generations, which contribute to the social cohesion of the community as well as to individual and collective freedom of choice and action.

Indigenous peoples

In September 2007, the United Nations General Assembly adopted the UN Declaration on the Rights of Indigenous Peoples (UNDRIP). Various UN and other development agencies, such as the United Nations Development Programme, the Food and Agriculture Organization, the International Fund for Agricultural Development and the World Bank, have adopted policies and operational guidelines on engaging with indigenous peoples. These reflect advancing standards with respect to indigenous peoples, strengthen partnerships, ensure greater relevance and positive outcomes, and also safeguard against unintended adverse effects. In 2008, the UN Development Group (UNDG) issued Guidelines on Indigenous Peoples' Issues, as a complement to earlier guidance on the application of the Human Rights-Based Approach (HRBA) to development. UNESCO has similarly embarked on a process to develop a UNESCO-wide policy on engaging with indigenous peoples.

Benefits beyond borders

For all these reasons, World Heritage – and heritage in general – is crucial to sustainable development, the physical and spiritual well-being of communities and to the building of mutual understanding and peace. Sustainable development is a goal that, almost by definition, acquires its meaning at a scale which is often much larger than that of a World Heritage property, suggesting that World Heritage planning and management need to be more integrated in territorial and regional strategies. This is also the view taken within the Recommendation on Historic Urban Landscapes (HUL) adopted by the General Conference of UNESCO at its 36th session (Paris, 2011). If heritage is to be part of the economic future of communities then it needs to be part of their future identity, distinctiveness and points of difference in the global competition to attract people and investment. *World Heritage: Benefits Beyond Borders* seeks to contribute to this larger goal by showcasing evidence-based knowledge drawn from a diverse range of World Heritage properties from around the world. A plethora of approaches and demonstration projects have been described through the case studies, which offer interesting and relevant experiences to share and lessons to learn.

One of the key learning outcomes from the case studies reviewed here is that we need to build the capacity of our partners and site managers to develop appropriate qualitative and quantitative measures for evidence-based claims for community benefits. The conventional cost benefit analysis could be furthered with benefit-sharing paradigms. Contingency valuation and choice modelling have long been used for valuing the environment. More recently some of these methods are also being used in the humanities and social sciences. Understanding the community and environmental benefits from the conservation of a World Heritage site gives added weight to the core concept of outstanding universal value.

Another recurring point is the value and role of responsible tourism in the conservation and promotion of World Heritage sites. The case studies of Australia's Great Barrier Reef and Viet Nam's Hoi An Ancient Town illustrate that measuring benefits from tourism could catalyse further investment in conservation. It is also evident that strategic interventions such as Egypt's Dahshur programme could establish pilot approaches to poverty alleviation and local job creation.

In soliciting and sourcing these case studies one of the challenges has been to engage the voices of the local experts and managers to contribute. As mentioned in the Introduction to this volume, the insider inputs into the discourse of World Heritage are crucial to promote relevance, cultural diversity and engagement. Having said this, we are conscious that many sites, especially those in the least developed countries and small island developing states, continue to face significant limitations of expertise and resources.

In conclusion, I recommend everyone to consider the World Heritage Convention with reference to the critical report produced by the UN task team on the post-2015 development agenda in the lead-up to the Rio+20 deliberations: *The Future We Want*, June 2012.[12] The Convention has the potential to be one of the flagship drivers of this agenda, within the framework of commitment to the culture in development agenda of UNESCO.

[12] http://www.uncsd2012.org/thefuturewewant.html/

Bibliography

Aas, C., Ladkin, A. and Fletcher, J. 2005. Stakeholder collaboration and heritage management. *Annals of Tourism Research*, Vol. 32, No. 1, pp. 28–48.

Access Economics Pty Ltd. 2008. Economic Contribution of the GBRMP, 2006–07. Report prepared for the Great Barrier Reef Marine Park Authority. Canberra, Australia.

Africa 2009. Quels argumentaires pour le patrimoine africain? Report of thematic seminar held at Ségou, Mali, March 2001.

APSARA homepage. Phnom Penh, Authority for the Protection and Management of Angkor and the Region of Siem Reap. http://www.autoriteapsara.org/en/apsara.html

APSARA/Fraser Thomas Ltd, in association with Boffa Miskell Ltd and Tourism Resource Consultants. 2007. Angkor Management Plan. Auckland, New Zealand.

Assembly of First Nations. 2005. *Getting from the Roundtable to Results.* Canada – Aboriginal Peoples Roundtable Process April 2004 – March 2005 Summary Report.

Attorre, F., Francesconi, N., Taleb, N., Scholte, P., Saeed, A., Alfo, M. and Bruno, F. 2007. Will dragonblood survive the next period of climate change? Current and future potential distribution of *Dracaena cinnabari* (Socotra, Yemen). *Biological Conservation*, Vol. 138, pp. 430–39.

Ballard, C. and Wilson, M. 2012. Unseen monuments; managing Melanesian cultural landscapes. In: J. Lennon and K. Taylor (eds), *Managing Cultural Landscapes*. Oxford, UK, Routledge, pp. 131–53.

Banfield, L., Van Damme, K. and Miller, A. 2011. *Evolution and Biogeography of the Flora of the Socotra Archipelago (Yemen).* In: D. Bramwell and J. Caujapé-Castells (eds), *The Biology of Island Floras*. Cambridge, UK, Cambridge University Press, pp. 197–225.

Bedaux, R. M. A. 1988. Tellem and Dogon material culture. *African Arts*, Vol. 21, No. 4, pp. 38–45.

Berkes, F., Folke, C. and Colding, J. (eds). 1998. *Linking Social and Ecological Systems:* Management Practices and Social Mechanisms for Building Resilience. Cambridge, UK, Cambridge University Press.

Bigio, A. G. 2010. *The Sustainability of Urban Heritage Preservation. The Case of Marrakesh.* Washington, D.C., Inter-American Development Bank.

Blaenavon World Heritage Site Partnership. 2005–12. *Blaenavon Heritage News*. Wales, Torfaen County Borough Council.

——— 2010. Blaenavon World Heritage Site Management Plan 2011–2016. Wales, Torfaen County Borough Council.

Boege, E. 2002. Protegiendo lo nuestro: Manual para la gestión ambiental comunitaria, uso y conservación de la biodiversidad de los campesinos indígenas de América Latina. Mexico City, INI Fondo para el Desarrollo de los Pueblos Indígenas de América Latina y el Caribe.

Boege, E. with G. Vidrales Chan et al. 2008. *El patrimonio biocultural de los pueblos indígenas de México*. Mexico City, Instituto Nacional de Antropología e Historia, Comisión Nacional para el Desarrollo de los Pueblos Indígenas.

Bokova, I. 2012. Address by Irina Bokova, Director-General of UNESCO on the occasion of the Launch Ceremony for the 40th Anniversary of Celebrations of the World Heritage Convention in Japan, Tokyo, 13 February 2012. (DG/2012/020.) unesdoc.unesco.org/images/0021/002152/215290e.pdf

Bonfil Batalla, G. 1994. *México profundo. Una civilización negada*. Mexico City, Grijalbo, Chap. 2.

Borrini-Feyerabend, G. and Hamerlynck, O. 2011. *Réserve de Biosphère Transfrontière du Delta du Sénégal – Proposition de Gouvernance Partagée*, in collaboration with Christian Chatelain and the Team Moteur de la Gouvernance Partagée des Aires Marines Protégées en Afrique de l'Ouest. International Union for Conservation of Nature–Commission on Environmental, Economic and Social Policy (IUCN–CEESP).

Brown, J., Currea, A.M., and Hay-Edie, T. 2010. COMPACT: Engaging local communities in stewardship of globally significant protected areas. UNDP/GEF Small Grants Programme. New York, NY, USA.

Burbridge, P. 2000. *The Nomination of the Wadden Sea Conservation Area as a World Heritage Site*. A Feasibility Study for the Trilateral Wadden Sea Cooperation/Common Wadden Sea Secretariat.

Cardiff, S. 2011. Aboriginal reconnection and reconciliation Jasper National Park of Canada. Presentation for George Wright Society Conference, New Orleans, March 2011.

Cheung, C. and DeVantier, L. 2006. *Socotra, a Natural History of the Islands and their People*. K. Van Damme, science editor, Odyssey Books and Guides. Hong Kong, Airphoto International Ltd.

Ciarcia, G. 2001. Dogons et Dogon. Retours au 'pays du réel'. *L'Homme*, Vol. 157, pp. 217–30.

Cissé, L., Dembélé, A., Cornet, L. and Joffroy, T. 2011. Inondation de Bandiagara de juillet 2007, Aide à la reconstruction de logements, Rapport Final. Misereor, Fondation Abbé Pierre.

Cissé, L., Guindo, P. and Joffroy, T. 2010. Le Temple d'Arou, falaises de Bandiagara, pays dogon. (Brochure.)

Cissé, L. and Joffroy, T. 2006. Falaises de Bandiagara, Pays Dogon, Plan de conservation et de gestion, Ministère de la Culture du Mali/World Monuments Fund.

Cissé, L., Joffroy, T. and Garnier P. (eds). 2010. *Recommandations pour la construction d'écoles en pays dogon.* (Guide.)

Conklin, H. C. 1980. *Ethnographic Atlas of Ifugao: A Study of Environment, Culture, and Society in Northern Luzon.* New Haven, Conn., Yale University Press.

Cooper, D. 2012. iSimangaliso – A new chapter in South Africa's conservation history. *Bonjanala.* Pretoria, National Department of Tourism, May–February.

CWSS. 2008. Nomination of the Dutch-German Wadden Sea as World Heritage Site. Wadden Sea Ecosystem No. 24. Wilhelmshaven, Germany, Common Wadden Sea Secretariat, World Heritage Nomination Project Group.

——. 2010. Sylt Declaration. Ministerial Council Declaration of the Eleventh Trilateral Governmental Conference on the Protection of the Wadden Sea. Wilhelmshaven, Germany, Common Wadden Sea Secretariat.

Day, J. C. 2011. Protecting Australia's Great Barrier Reef. *Solutions*, Vol. 2, No. 1, pp. 56–66. http://www.thesolutionsjournal.com/node/846

DCMS. 1999. *World Heritage Sites: The Tentative List of the United Kingdom of Great Britain and Northern Ireland.* London, UK Government Department for Culture, Media and Sport.

De Geest, P. (ed.). 2006. Soqotra Karst Project (Yemen), 2000–2004. *Berliner Höhlenkundliche Berichte*, Vol. 20.

Deidun, A. 2004. Challenges to the conservation of biodiversity of small islands: the case of the Maltese Islands. *International Journal of Arts and Sciences*, Vol. 3, pp. 175–87.

Doquet, A. 1999. *Les masques dogon. Ethnologie savante, ethnologie autochtone.* Paris, Kharthala.

——. 2006. Décentralisation et reformulation des traditions en Pays dogon : les manifestations culturelles des communes de Dourou et Sangha. In: C. Fay, F. Koné and C. Quiminal (eds), *Décentralisation et pouvoirs en Afrique. En contrepoint, modèles territoriaux français.* Paris, Éditions de l'IRD. (Colloques et séminaires.)

Dulawan, L. 2001. *Ifugao Culture and History.* Manila, National Commission for Culture and the Arts.

ECOTEC Research and Consulting Ltd. 2010. *Valuing the Welsh Historic Environment.* UK National Trust.

El Faïz, M. 2002. *Marrakech patrimoine en péril.* Casablanca, Actes Sud, Eddif.

Elie, S. D. 2004. Hadiboh: from peripheral village to emerging city. *Chroniques Yéménites*, Vol. 12, pp. 53–80.

——. 2008. The waning of Soqotra's pastoral community: political incorporation as social transformation. *Human Organization*, Vol. 67, pp. 335–45.

Escher, A., Petermann, S. and Clos, B. 1999. Gentrification in der Medina von Marrakech. *Geographische Rundschau*, Vol. 53, No. 6, 1999, pp. 24–31.

Escher, E. 2000. Le bradage de la médina de Marrakech? In: *Le Maroc à la veille du Troisième Millénaire. Défis, chances et risques d'un développement durable, Colloques et Séminaires*, 93. Rabat, Faculté des Lettres et des Sciences Humaines, pp. 217–32.

Freire, P. 2005. *Pedagogía del Oprimido*. Mexico City, Ed. Siglo XXI, 2nd edn in Spanish, Chap. 2.

GBRMPA You-Tube Channel <http://www.youtube.com/user/TheGBRMPA>; specifically videos in the folder 'Reef Guardians Stewardship Program 2011' and 'Working with Communities and Industries'

GBR *Outlook Report 2009*: http://www.gbrmpa.gov.au/outlook-for-the-reef/great-barrier-reef-outlook-report

GBR rezoning – first public participation phase: http://www.gbrmpa.gov.au/zoning-permits-and-plans/rap/first-community-participation-phase

Getty Publications. 2008. Proceedings Terra 2008: the 10th International Conference on the Study and Conservation of Earthen Architectural Heritage, organized by the Getty Conservation Institute and the Malian Ministry of Culture, Bamako, Mali, 1–5 February.

Government of Cambodia. 2004. Decision on the Standards of Land Use in the Protected Zones 1 and 2 of the Angkor Park (Phnom Penh, 16 November, Decision No. 70 SSR).

Great Barrier Reef World Heritage Area: http://www.gbrmpa.gov.au/about-the-reef/heritage/great-barrier-reef-world-heritage-area

Guertin, P. 2008. Run Ta-Ek. The creation of a US ecovillage in Khmer traditional medium. Phase II: principles and the central village development plan. Siem Reap/Quebec, APSARA.

Hay-Edie, T. and Brown, J. 2010. COMPACT: Engaging local communities in stewardship of globally significant protected areas. New York, UNDP/GEF Small Grants Programme.

Hay-Edie, T., Murusuri, N., Moure, J. and Brown, J. 2012. Engaging local communities in World Heritage sites: experience from the Community Management for Protected Areas Programme. In: S. Weber (ed.), *Rethinking Protected Areas in a Changing World*. Proceedings of the 2011 George Wright Society Biennial Conference on Parks, Protected Areas, and Cultural Sites.

Hu Shujiong. 2002. *Kaiping Diaolou*, trans. S. Xie. Beijing, Zhongguo she ying chu ban she.

Ifugao Cultural Heritage Office. 2011. Information on the List of Rice Terrace Organizations from the Ifugao Province (as of December 2011).

iSimangaliso Wetland Park Authority. 2005. Craft Programme Progress Report. St Lucia, iSimangaliso Wetland Park Authority.

2008. *iSimangaliso News*, Vol. 1, No. 3, p. 2.

2010. iSimangaliso Wetland Park Authority, People and Parks Progress Report, April 2009 to March 2010, submitted to the Department of Environment, Pretoria.

2011. *iSimangaliso Wetland Park Integrated Management Plan (2011–2016)*, St Lucia, iSimangaliso Wetland Park Authority.

2012. *About iSimangaliso*. St Lucia, iSimangaliso Wetland Park Authority.

IUCN. 1999. World Heritage Nomination: IUCN Technical Evaluation: Greater St Lucia Wetland Park (South Africa). Gland, Switzerland, International Union for Conservation of Nature. http://whc.unesco.org-archive-advisory_body_evaluation-914.pdf

IUCN/UNESCO 2008. *Socotra Archipelago* – World Heritage Nomination – IUCN Technical Evaluation (Yemen). International Union for Conservation of Nature Evaluation Report 1263, May. Paris, United Nations Educational, Scientific and Cultural Organization.

Job, H. 2009. Regionalökonomische Effekte des Tourismus in deutschen Nationalparken. *Naturschutz und Biologische Vielfalt*, Bonn, Vol. 76.

Jokilehto, J. 2006. World Heritage: Defining the outstanding universal value. *City and Time*, Vol. 2, No. 2, p. 1. http://www. ct.ceci-br.org

Khuon, K.-N. 2005. A commitment to community engagement. In: *Phnom Bakheng Workshop on Public Interpretation, Angkor Park, Siem Reap, Cambodia*. Siem Reap, APSARA/World Monuments Fund/Center for Khmer Studies, pp. 115–17.

—— 2006. Angkor site management and local communities. Paper presented at the conference Angkor, Landscapes, City and Temples, University of Sydney, Australia.

—— 2008. Run Ta-Ek. Eco-village for sustainable development. Concept note of an ecological human settlement. Phnom Penh/Siem Reap, APSARA.

—— 2009*a*. *Archaeological Park and Angkor Ceramic Museum at Tani*. Phnom Penh/Siem Reap, APSARA.

—— 2009*b*. Maximizing community benefits from Angkor heritage. Paper presented to Advanced Sustainable Tourism at Cultural and National Heritage Site Workshop, Dunhang, China.

Kingdom of Cambodia/APSARA/UNESCO. 1996. *Angkor – A Manual for the Past, Present and Future*. Compiled by A. Chouléan, E. Prenowitz and A. Thompson under the supervision of Vann Molyvann, Minister of State for Culture and Fine Arts, Territorial Management, Urban Planning and Construction. Phnom Penh, APSARA.

Klopper, S. 1992. The art of Zulu-Speakers in Northern Natal-Zululand: an investigation of the history of beadwork, carving and dress from Shaka to Inkatha. Ph.D. thesis, University of Witwatersrand, Johannesburg.

Knapp, R. G. 2005. *Chinese Houses: The Architectural Heritage of A Nation*. North Clarendon, Vt., Tuttle Publishing.

Kruger, F. J., van Wilgen, B. W., Weaver, A. V. B. and Greyling, T. 1997. Sustainable development and the environment: lessons from the St Lucia Environmental Impact Assessment. *South African Journal of Science*, Vol. 93, November, pp. 23–33.

Kurzac-Souali, A.-C. 2006. Les médinas marocaines : une requalification sélective. Élites, patrimoine et mondialisation au Maroc. Doctoral thesis, Université Paris IV Sorbonne.

Langdon, S., Prosper, R. and Gagnon, N. 2010. Two paths one direction: Parks Canada and Aboriginal peoples working together. *The George Wright Forum*, Vol. 27, No. 2, pp. 222–33.

Lazcano-Barrero, M. A. 1990. Conservación del cocodrilo en Sian Ka'an. *Amigos de Sian Ka'an*, Bulletin No. 5, Cancún, Quintana Roo, Mexico.

Living with Heritage (LWH). University of Sydney, Australia. http://www.acl.arts.usyd.edu.au/angkor/lwh/index.php

Local Marine Advisory Committees: http://www.gbrmpa.gov.au/our-partners/local-marine-advisory-committees

López-Ornat, A. 1990. Avifauna de la Reserva de la Biosfera de Sian Ka'an. In: *Diversidad biológica en la Reserva de la Biosfera de Sian Ka'an, Quintana Roo, México.* Mexico, Centro de Investigaciones de Quintana Roo, pp. 332–70.

Lung, D. P. Y. 1991. *Chinese Traditional Vernacular Architecture.* Hong Kong Regional Council and University of Hong Kong.

MacKinnon, H. B. 1992. Field check-list of the birds of the Yucatán peninsula and its protected areas. *Amigos de Sian Ka'an,* Cancún, Quintana Roo, Mexico.

MacLaren, I. S. 2007 (ed.). *Culturing Wilderness in Jasper National Park. Studies in Two Centuries of Human History in the Upper Athabasca River Watershed.* Edmonton, Alta, University of Alberta Press.

Madulid, D. 2010. *Plant Diversity and Conservation of the Woodlots of Ifugao.* Pasay City, UNESCO National Commission of the Philippines.

Makino, M., Matsuda, H. and Sakurai, Y. 2009. Expanding fisheries co-management to ecosystem-based management: a case in the Shiretoko World Natural Heritage area, Japan. *Marine Policy,* Vol. 33, pp. 207–14.

Mananghaya, B. J. *Heritage: Driver for Development and the Case of the Rice Terraces of the Philippine Cordilleras.* Paper presented at the 17th ICOMOS General Assembly, 26 November to 3 December 2011, Paris.

Mayer, A.-N. 2009. Probleme touristischer Entwicklung auf der Insel Soqotra. Vom Missverständnis 'Ökotourismus' zu nachhaltigem Tourismus? *Jemen Studien,* Vol. 19, pp. 1–152.

Miller, M. and Morris, M. 2004. *Ethnoflora of the Soqotra Archipelago.* Edinburgh, UK, Royal Botanical Gardens.

Mngomezulu, Z. A. 2009. A qualitative analysis of the economic benefits of the Land Care Project on rural SMME contractors living around the iSimangaliso Wetland Park. Honours thesis, geography and environmental management, University of KwaZulu Natal, Durban.

Moon, O. 2005. Hahoe: the appropriation and marketing of local cultural heritage in Korea. *Asian Anthropology,* Vol. 4, pp. 1–29.

Morris, M. 2002. Manual of traditional land use in the Soqotra Archipelago. Edinburgh, UK, Royal Botanical Garden. (Unpublished report, GEF/YEM/96/G32.)

Mothana, R. A. A., Lindequist, U., Gruenert, R. and Bednarski, P. J. 2009. Studies of the in vitro anticancer, antimicrobial and antioxidant potentials of selected Yemeni medicinal plants from the island. *BMC Complementary and Alternative Medicine,* Vol. 9, No. 7.

Ngcobo, P. N. 2009. The socio-economic impacts of the iSimangaliso Wetland Park Land Care Programme on contract workers. Unpublished honours thesis, geography and environmental management, University of KwaZulu Natal, Durban.

NIKE Project Reports. 2009. Manila, Ifugao State University, Nurturing Indigenous Knowledge Experts.

Nyamweru, C. K. 1998. *Sacred Groves and Environmental Conservation*. Frank P. Piskor Lecture. New York, St Lawrence University, pp. 2–27.

Parks Canada Agency. 2004. Periodic Report on the Application of the World Heritage Convention. Report on the State of Conservation of Canadian Rocky Mountain Parks. http://www.pc.gc.ca/eng/docs/rspm-whsr/rapports-reports/r1.aspx

— 2010. Jasper National Park of Canada Management Plan. Jasper, Alta, Parks Canada Agency.

— 2011*a*. *A Handbook for Parks Canada Employees on Consulting and Accommodation with Aboriginal Peoples*. Ottawa, Ont., Aboriginal Affairs Secretariat.

— 2011*b*. *Working Together: Our Stories. Best Practices and Lessons Learned in Aboriginal Engagement*. Gatineau, PQ, Aboriginal Affairs Secretariat. aboriginal.autochtones@pc.gc.ca

Peutz, N. 2011. 'Shall I tell you what Soqotra once was?' World Heritage and sovereign nostalgia in Yemen's Soqotra Archipelago. *Transcontinentales*, Vol. 10/11, pp. 1–14.

— 2012. Revolution in Socotra. A perspective from Yemen's periphery. *Middle East Report*, Vol. 42, No. 2, pp. 14–20.

Pietsch, D. and Morris, M. 2010. Modern and ancient knowledge of conserving soils in Socotra Island, Yemen. In: P. Zdruli, M. Pagliai, S. Kapur and A. Faz Cano (eds), *Land Degradation and Desertification: Assessment, Mitigation and Remediation*. Dordrecht, Netherlands, Springer Publishing, pp. 374–86.

Platzky, L. and Walker, C. 1985. *The Surplus People: Forced Removals in South Africa*. Johannesburg, Ravan. (For the Surplus People Project.)

Porter, R., Swift, M., James, B., Scott, D., Zaloumis, A., Makgolo, M. and Palmer, G. 2003. World Heritage: the South African experience. In: G. Cowan, J. Yawitch and M. Swift (eds), *Strategic Innovations in Biodiversity Conservation: The South African Experience*. Pretoria, Department of Environmental Affairs and Tourism, pp. 73–95.

Prognos AG. 2004. Sector Specific Analysis and Perspectives for the Wadden Sea Region. (Wadden Sea Forum Report No. 8.)

Province of Ifugao/UNESCO National Commission of the Philippines/J.B Mananghaya. *The 2008 State of Conservation Report on the Rice Terraces of the Philippine Cordilleras*, submitted by the National Commission to the World Heritage Centre. 2009, 2010, 2011, 2012 State of Conservation Reports on the Rice Terraces of the Philippine Cordilleras.

Rebanks Consulting Ltd/Trends Business Research Ltd.

— 2010. *World Heritage: Is there Opportunity for Economic Gain?* Research and Analysis of the Socio Economic Impact Potential of UNESCO World Heritage Site Status: http://www.lakeswhs.co.uk/documents/WHSTheEconomicGainFinalReport.pdf

Reef Guardians: http://www.gbrmpa.gov.au/our-partners/reef-guardians

Regenvanu, R. 2008. Keynote Address, International Workshop on Traditional Knowledge Systems, Museums & Intangible Natural Heritage in South Asia. ICOM and Salarjung Museum Hyderabad, 3–7 February.

Reser, J. P. and Bentrupperbaumer, J. M. 2005. What and where are environmental values? Assessing the impacts of current diversity of use of 'environmental' and 'World Heritage' values. *Journal of Environmental Psychology*, Vol. 25, pp. 125–46.

Royal Decrees. *See* references listed under Sihanouk.

Rypkema, D. D. 2005. *Cultural Heritage and Sustainable Economic and Social Development*. European Cultural Heritage Forum, Brussels.

Saïgh Bousta, R. 2004. Le Ryad maison-d'hôte, esquisse d'une réflexion sur le phénomène et ses retombées. In: *Le Tourisme durable: quelles réalités et quelles perspectives*. Marrakech, Université Cadi Ayyad, Faculté des Lettres et des Sciences Humaines, pp. 157–69.

Scholte, P. and De Geest, P. 2010. The climate of Socotra Island (Yemen): a first-time assessment of the timing of the monsoon wind reversal and its influence on precipitation and vegetation patterns. *Journal of Arid Environments*, Vol. 74, pp. 1407–1515.

Scholte, P., Al-Okaishi, A. and Suleyman, A. S. 2011. When conservation precedes development: a case study of the opening of the Socotra archipelago, Yemen. *Oryx*, Vol. 45, No. 3, pp. 401–10.

Sihanouk, N. 1994. Royal Decree establishing Protected Cultural Zones in the Siem Reap/Angkor Region and Guidelines for their Management (Phnom Penh, 28 May 1994, Decree 001/NS). Phnom Penh, APSARA. http://www.autoriteapsara.org/en/apsara/about_apsara/legal_texts/decree3_text.html

—— 1995. Royal Decree establishing a National Authority for the Protection and Management of Angkor and the Region of Siem Reap, named APSARA (Phnom Penh, 19 February 1995, Decree NS/RKT/0295/12). Phnom Penh, APSARA. http://www.autoriteapsara.org/en/apsara/about_apsara/legal_texts/decree1_text.html

—— 1996. Government of Cambodia. Law on the Protection of Cultural Heritage. Phnom Penh, 25 January 1996, Decree NS/RKM/0196/26). Phnom Penh, APSARA. http://www.autoriteapsara.org/en/APSARA/about_apsara/legal_texts/decree4_text.html

Skounti, A. 2004. Marrakech: patrimoine versus «élitisation». Processus de patrimonialisation, pauvreté et gestion de la médina. In: *Patrimoine et développement durable dans les villes historiques du Maghreb. Enjeux, diagnostics et recommandations*. Rabat, UNESCO, pp. 143–56.

—— 2009. The authentic illusion: humanity's intangible cultural heritage, the Moroccan experience. In: L. Smith and N. Akagawa (eds), *Intangible Heritage*, London, Routledge, pp. 74–92.

—— 2011. The lost ring. The UNESCO World Heritage and Intangible Cultural Heritage. *Milli Folklor (Turkey)*, Vol. 23, No. 89, pp. 28–40.

Sohlman, E. 2004. A bid to save 'the Galápagos of the Indian Ocean'. *Science*, Vol. 303, p. 1753.

Souza, A. I. 2011. *Paulo Freire, vida e obra*. São Paulo, Brazil, Ed. Expressao Popular.

Spear, T. T. 1978. *The Kaya Complex: A History of the Mijikenda Peoples of the Kenya Coast to 1900*. Nairobi, Kenya Literature Bureau.

Taamouti, M. et al. 2008. Médinas marocaines. *Les Cahiers du Plan*, No. 20, September-October. Rabat, Publications du Haut Commissariat au Plan.

Tebbaa, O. 2010. Patrimoine, patrimonialisation et développement touristique: le cas de Marrakech. Hespéris-Tamuda. *Revue de la Faculté des Lettres et des Sciences Humaines*, Rabat, Vol. XLV, pp. 55–66.

Thorsby, D. 2008. *Culture in Sustainable Development: Insights for the Future Implementation of Art. 13 [Convention on the Protection and Promotion of the Diversity of Cultural Expressions]*. http://unesdoc.unesco.org/images/0015/001572/157287e.pdf

Tolba, M. 2004. Environment and development in the approaches to 2020. In: J. Bindé, *The Future of Values: 21st-Century Talks*. Paris/Oxford, UK, UNESCO/Berghahn, pp. 210–14.

Toledo, V M. and Barrera-Bassols, N. 2008. *La memoria biocultural (la importancia ecológica de las sabidurías tradicionales)*. Barcelona, Spain, Icaria Editorial.

Tollrianova, Z. 2011. Turtle hunting continues in reserve. *Tayf (Friends of Soqotra Newsletter)*, Vol. 8, p. 13.

Toufiq, A. 1988. Haoula Maâna Ism Murrakuch [On the meaning of Marrakesh]. In: *Murrakuch. Min Al-Ta'sis Ila Akhir Al-Asr Al-Muwahhidi* [Marrakesh. From the Foundation to the End of the Almohad Reign]. Casablanca, Université Cadi Ayyad de Marrakech, Imprimerie Fedala, pp. 15–19.

Traditional Owners: http://www.gbrmpa.gov.au/our-partners/traditional-owners.

Trau, A. M. 2012. Beyond pro-poor tourism: (re)interpreting tourism-based approaches to poverty alleviation in Vanuatu. *Tourism Planning & Development*, Vol. 9, No. 2, pp. 149–64.

UNDP. 2003. *Sustainable Development and Biodiversity Conservation for the People of Socotra Islands, Yemen*. New York, United Nations Development Programme. (Project Document YEM/03/004/A/01/99.)

___ 2008. *Strengthening Socotra's Policy and Regulatory Framework for Mainstreaming Biodiversity*. New York, United Nations Development Programme. (Project Document, Atlas Award ID: 00049646.)

UNEP/WCMC. 2008. *Socotra Archipelago, Yemen*. UNESCO Fact Sheet. World Heritage sites – Protected Areas and World Heritage. United Nations Environment Programme/World Conservation Monitoring Centre. www.unep-wcmc.org/sites/wh/pdf/Socotra%20revised.pdf

UNESCO Mexico. 2009. *Estudio de la contribución de los Sitios de Patrimonio Mundial al Desarrollo*. Final Report, November. Mexico City, United Nations Educational, Scientific and Cultural Organization.

UNESCO. 1972. Convention concerning the Protection of the World Cultural and Natural Heritage, adopted by the General Conference at its 17th Session, 16 November. Paris, UNESCO Headquarters. http://whc.unesco.org-archive-advisory_body_evaluation-914.pdf

___ 1992. World Heritage Committee 16th session, Santa Fe, United States (7–11 December 1992). Report [Decision on the NSI AFF of Angkor on the UNESCO World Heritage List: 40–41]. Paris, UNESCO. English version (including Justification for Inscription on the List of World Heritage in Danger): http://WHC.UNESCO.org/archive/repcom92.htm#Angkor

1993. Inter-governmental conference on the safeguarding and development of the historical area of Angkor (Tokyo 12–13 October 1993). Paris, UNESCO Archive (unpublished document).

2003. Paris Declaration – Safeguarding and Development of Angkor (November 15, 2003). Paris, UNESCO (Document CLT/2003/ME/H/1.) http://unesdoc.unesco.org/images/0015/001588/158898e.pdf

2006. *The UNESCO World Heritage Centre's Natural Heritage Strategy*. Paris, World Heritage Centre. (CLT.2006/ws/12) http://whc.unesco.org/uploads/activities/documents/activity-398-1.pdf

2011. Convention concerning the Protection of the World Cultural and Natural Heritage, World Heritage Committee, Thirty-fifth Session, 19–29 June. Paris, UNESCO Headquarters. (WHC-10/35.COM/12D.) whc.unesco.org/document/106392

Van Beek, W. E. A. 1991. Dogon restudied: a field evaluation of the work of Marcel Griaule. *Current Anthropology*, Vol. 32, No. 3, pp. 139–67.

Van Damme, K. 2009. *Socotra Archipelago*. In: R. G. Gillespie and D. A. Clague (eds), *Encyclopedia of Islands*. Encyclopedias of the Natural World, University of California Press, pp. 846–51.

2011. Insular biodiversity in a changing world. *Nature Middle East*, doi:10.1038/nmiddleeast.2011.61, published online 25 May 2011.

Van Damme, K. and Banfield, L. 2011. Past and present human impacts on the biodiversity of Socotra Island (Yemen): implications for future conservation. *Zool. Middle East*, Suppl. 3, pp. 31–88.

Walker, C. 2008. *Landmarked: Land Claims and Land Restitution in South Africa*. Jacana/Ohio University Press, Johannesburg/Athens, USA.

World Heritage Committee. 2009*a*. Decision 33COM 8B.4. Paris, UNESCO World Heritage Centre.

2009*b*. Document WHC-09/33.COM/INF.8B2. Paris, UNESCO World Heritage Centre.

Yahia, M. 2011. Foreign researchers flee Yemen leaving conservation programmes in trouble. *Nature Middle East*, doi:10.1038/nmiddleeast.2011.36, published online 22 March 2011.

Zajonz, U., Aideed, M. S., Saeed, F. N., Lavergne, E., Klaus, R. and Krupp, F. 2012*a*. Sustainable traditional fisheries management on Socotra: a tale of wishful thinking? *Tayf (Friends of Soqotra Newsletter)*, Vol. 9.

Zajonz, U., Klaus, R., Abdul-Assiz, M. and Saeed, F. N. 2012*b*. Coral mining on Soqotra – little profit for a few and a lot of damage for many. *Tayf (Friends of Soqotra Newsletter)*, Vol. 9.

Zajonz, U., Klaus, R., Pulch, H., Lavergne, E., Naseeb, F. N., Ziegler, M., Alpermann, T., Goerres, M. and Krupp, F. 2011. Socotra Research Projects. Summary Progress Report 2009–2010 to the Environment Protection Authority and the Ministry of Water and Environment of the Republic of Yemen. Frankfurt am

Main, Germany, Tropical Marine Ecosystems Group Biodiversität und Klima Forschungszentrum und Senckenberg Forschungsinstitut. (Unpublished report.)

Zaloumis, A. P. 2005. South Africa's spatial development initiatives: the case study of the Lubombo SDI and the Greater St Lucia Wetland Park. Master's thesis, University of KwaZulu-Natal, Durban.

Contributors

Abungu, George H.O.

Ph.D., Cambridge-trained archaeologist and former Director-General of the National Museums of Kenya, he is the founding chairman of Africa 2009, ISCOTIA (International Standing Committee on the Traffic in Illicit Antiquities), and Vice-President of ICOM. He was Kenya's representative to the UNESCO World Heritage Committee and Vice-President of its Bureau (2004–09).

Alatalu, Riin

Holder of double Master's degree in restoration and conservation from the Estonian Art Academy and History from Tartu University, Head of the Division of Milieu Areas of Tallinn Culture and Heritage Department, President of ICOMOS Estonia, she has published several articles and lectured on World Heritage policy and international organizations.

Ball, Cynthia

As an Environmental Assessment Specialist with Parks Canada, her work encompasses a broad range of project assessments and restoration initiatives that contribute to maintaining and improving the condition of Jasper National Park and the Canadian Rocky Mountain Parks World Heritage site.

Brown, Jessica

A global consultant with the Community Management of Protected Areas Conservation Programme (COMPACT) initiative, she is Executive Director of the New England Biolabs Foundation and chairs the IUCN-WCPA Specialist Group on Protected Landscapes. She holds degrees from Clark University and Brown University and has worked in countries of Latin America, the Caribbean and Central and Eastern Europe.

Cardiff, Shawn

As Manager of Land Use Policy and Planning, he is broadly involved in the protection, presentation, reporting, and experiences offered to visitors of Jasper National Park and the Canadian Rocky Mountain Parks World Heritage site, and has also worked at Canada's Nahanni National Park World Heritage site.

Cissé, Lassana

Master of Philosophy in cultural anthropology, Director of the Cultural Mission and Manager of the Cliff of Bandiagara (Land of the Dogons) World Heritage site (Mali). Since 1994, he has worked to implement several projects linking heritage values conservation and development in Dogon Country, involving local communities, stakeholders and traditional chiefs.

Day, Jon

As Director for Planning, Heritage and Sustainable Funding within the Great Barrier Reef Marine Park Authority, he has undertaken a variety of planning and management roles in the Great Barrier Reef World Heritage Area since 1986. He has worked on several other Australian World Heritage properties, helped to develop the current Periodic Reporting process, and represented Australia on the World Heritage Committee.

Debevec, Vanja

Board member of the Permanent Commission for Speleotherapy of the International Union of Speleology, and its Secretary General from 1997 to 2008, She started working in 1999 in the Škocjan Caves park (Slovenia), as Head of Department for research and development.

Engels, Barbara

Holder of a diploma in biology and a Master's degree in European studies, she is a World Heritage specialist at the German Federal Agency for Nature Conservation. As a member of IUCN-WCPA she is working on natural World Heritage including serial nominations, sustainable tourism and capacity-building.

Galla, Amareswar

Ph.D., Alumnus of Jawaharlal Nehru University, New Delhi, professor and Executive Director, International Institute for the Inclusive Museum, Copenhagen, he has worked on culture in poverty alleviation projects at several World Heritage sites including Ha Long Bay and Hoi An (Viet Nam) and Darjeeling Himalayan Railway (India).

Githitho, Anthony Ngacha

Senior research scientist with the National Museums of Kenya, he heads the Coastal Forest Conservation Unit which is primarily concerned with conservation of the Mijikenda Kaya Sacred Forests of Coastal Kenya in collaboration with local communities, as well as biodiversity research.

Govender, Nerosha

Research officer at iSimangaliso Authority managing research, environmental education and awareness programmes and coordinating UNESCO and transfrontier conservation activities. She graduated with honours in geography and environmental management and is currently finalizing her Master's thesis which looks at tourism shifts in the iSimangaliso Wetland Park (South Africa) since gaining World Heritage status.

Guidon, Niéde
Ph.D. from the École de Hautes Études en Sciences Sociales (Paris), is professor and President of the Foundation Museum of the American Man and researcher at the National Institute of Archaeology, Palaeontology and Semi-arid Environment of Brazil. He has conducted research since 1973 in the Serra da Capivara National Park World Heritage site.

Hay-Edie, Terence
Programme advisor for the UNDP-implemented Global Environment Facility Small Grants Programme (operating in 125 countries), and Community Management of Protected Areas Conservation Programme (COMPACT) partnership focusing community-level small grants for nine World Heritage sites, he holds a Ph.D. from Cambridge University and post-doc (College de France) with extensive fieldwork conducted on World Heritage cultural landscapes in Nepal, China, Ecuador and Mali.

Inniss, Tara A.
Ph.D., currently lecturer in the Department of History and Philosophy at Cave Hill Campus, University of the West Indies (Barbados). The areas of focus for her teaching and research include history of medicine; history of social policy; heritage and social development.

James, Bronwyn
Master's degree in geography and environmental science, specializing in rural development and environment, has worked as researcher for a land rights NGO and then at the Energy and Development Research Centre at the University of Cape Town (South Africa). Having set up and implemented the craft and cultural tourism programmes of the Lubombo Spatial Development Initiative, she now heads the community development, capacity-building and research areas with the iSimangaliso Authority.

Johansen, Rita
Site coordinator of the Vega Archipelago (Norway) and Managing Director of the Vega World Heritage Foundation, responsible for local management and involvement, dissemination and information, she was one of the local enthusiasts taking the initiative to have the Vega Archipelago nominated for World Heritage status, both to protect its values and initiate local value creation.

Khuon, Khun-Neay
Deputy Director General of the National Authority for Protection and Management of Angkor World Heritage site (APSARA) in Cambodia, in charge of land and habitat management. Works on the Living with Heritage project with Australia,

347

the Run Ta-Ek Eco-Village project (APSARA) and the Angkor Participatory Natural Resource Management and Livelihoods project with New Zealand.

Londoño, Juan Luis Isaza

An architect with studies in Spanish-American art history, he has worked in different Colombian organizations related to culture, architecture and the preservation of cultural heritage. Consultant to international organizations such as ICOMOS, UNESCO and the World Monuments Fund, he is currently Heritage Director of the Colombian Ministry of Culture.

Makino, Mitsutaku

M.Phil. (Cambridge), M.A. and Ph.D. (Kyoto), specializes in marine resource management and environmental policy analysis. After working as a researcher in Yokohama National University, he is now the Head of Fisheries Management Group, National Research Institute of Fisheries Science, Fisheries Research Agency, Japan.

Mananghaya, Joycelyn

Alumnus of the University of the Philippines College of Architecture and the Escuela Nacional de Conservacion, Restauracion y Museografia, DF Mexico, Faculty and Dean of the College of Architecture, FEATI University, Manila. Recently she has worked on the removal of the Rice Terraces of the Philippine Cordilleras from the List of World Heritage in Danger.

Martin, Gabriela

Ph.D., professor and Deputy Head of the postgraduate degree course in archaeology at the Federal University of Pernambuco, Scientific Director of the Foundation Museum of the American Man and Deputy Head of the National Institute of Archaeology, Palaeontology and Semi-arid Environment of Brazil. Her research is on the prehistory of North-East Brazil, including the Serra da Capivara National Park World Heritage site.

Mbaye, Khatary

Local coordinator, Senegal, for the Community Management of Protected Areas Conservation Programme (COMPACT), implemented by the UNDP Global Environment Facility Small Grants Programme for the Djoudj-Djawling Transboundary Biosphere Reserve (Mauritania / Senegal). He has a background in participatory natural resource management and has worked in various protected areas across Senegal.

Miyazawa, Yasuko

Technical official, Ministry of the Environment, currently in charge of World Natural Heritage in Japan.

Meropoulis, Sherrill
As the Aboriginal Liaison Officer for Jasper National Park (Canada), she is building enduring working relationships with the many communities that have historical associations with Jasper National Park and the Canadian Rocky Mountain Parks World Heritage site.

Moure, Julio
Regional Coordinator, COMPACT-Mexico, was trained as an educator in Spain and Paris and has extensive field experience living and working in Mexico, Mozambique and Brazil. As COMPACT Coordinator he has spent the last twelve years working with Mayan communities in Quintana Roo (Mexico) in the landscape of Sian Ka'an World Heritage site.

Pessis, Anne-Marie
Ph.D., professor and Head of the postgraduate degree in archaeology at the Federal University of Pernambuco, researcher at the Foundation Museum of the American Man and Head of the National Institute of Archaeology, Palaeontology and Semi-arid Environment of Brazil. Her research is on the rock-art of North-East Brazil, mainly in the Serra da Capivara National Park World Heritage site.

Rao, Kishore
Director of UNESCO's World Heritage Centre, which is the Secretariat to the World Heritage Committee under the World Heritage Convention. He holds Master's degrees in resource policy and planning from Cornell University (United States) and in forestry from the Forest Research Institute in India. He has worked on World Heritage conservation issues over the past thirty years with the Government of India, with the International Union for Conservation of Nature (IUCN) and with UNESCO.

Scott, Dianne
Ph.D., honorary research fellow in the School of Built Environment and Development Studies, University of KwaZulu-Natal (South Africa), she is an environmental social scientist who has supervised postgraduates; worked as research and policy coordinator; cooperated in a knowledge-building project; and designed and delivered training modules for iSimangaliso Wetland Park.

Skounti, Ahmed
Anthropologist at the Institut National des Sciences de l'Archéologie et du Patrimoine (Rabat, Morocco) and associate professor at the University of Marrakesh. He holds a Ph.D. from the École des Hautes Etudes en Sciences Sociales, Paris. UNESCO consultant on issues relating to the World Heritage Convention (1972) and the Intangible Heritage Convention (2003), he participated in the drafting of the 2003 Convention.

Sow, Mamadou Samba
Local coordinator, Mauritania, for the Community Management of Protected Areas Conservation (COMPACT) Programme, implemented by the UNDP Global Environment Facility Small Grants Programme for the Djoudj-Djawling Transboundary Biosphere Reserve (Mauritania / Senegal). He has a background in forestry and GIS from the University of Fes and Rabat (Morocco).

Stewart, Amber
Holder of a Master's degree in planning from the University of Calgary, since 1997 she has worked in a variety of Parks Canada positions in the Canadian Rocky Mountain Parks World Heritage site and is currently the Land Use Planner for Jasper National Park of Canada.

Thomas, Catherine
Head of Regeneration Services for Torfaen County Borough Council (Wales, UK) with over thirty years experience in local government, her work over the past decade has focused on using Blaenavon's World Heritage status as a catalyst for the development of cultural tourism, educational opportunities and economic regeneration.

Van Damme, Kay
Ph.D., researcher, College of Life and Environmental Sciences, University of Birmingham (United Kingdom), he has been involved with the Socotra Archipelago (Yemen) since 1999. He chairs Friends of Soqotra, raising awareness on the Socotri culture and environment and stimulating projects that aim at the involvement of local communities and protection of the unique insular ecosystems and their people.

Vohland, Karen
Holder of a Master's degree in professional communications and Director of Stakeholder Engagement and Stewardship, Great Barrier Reef Marine Park Authority (Australia), she is a passionate communications professional with more than twenty-five years experience in media, public relations and stakeholder engagement having held senior roles in human services and environmental organizations.

Westrik, Carol
Holder of a Master's degree in landscape conservation and history of art and a Ph.D. in post-war reconstruction and development, she currently works as an advisor on heritage issues, with special emphasis on World Heritage.

Wren, Liz
Director of the Indigenous Partnerships Group at the Great Barrier Reef Marine Park Authority (Australia), she has worked extensively with Aboriginal groups across the country, collaborating with local communities to protect natural and cultural heritage that is important to them.

Yang, Minja

Alumnus of the University of London's School of Oriental and African Studies and Georgetown University (United States), she became President of Raymond Lemaire Centre for Conservation (Leuven, Belgium) since retiring from UNESCO. While serving as Deputy Director, World Heritage Centre; Director for Museums, then Director of the UNESCO New Delhi Office she focused on World Heritage Cities projects.

Zhan, Guo

Former counsel, senior researcher of China State Administration of Cultural Heritage; Vice President of ICOMOS; Vice President and Secretary General of ICOMOS China; Chairman of Chinese Expert Committee for World Cultural Heritage. He presided over the nomination and management of Word Heritage cultural sites and cultural parts of mixed heritage sites in China before 2009, with extensive involvement in World Heritage worldwide.

Photo Credits

p. 49 © MAJ Gabriela / SIPA.

p. 50 © Our Place World Heritage collection.

p. 51 © Our Place World Heritage collection.

5   p. 54 © Inge Ove Tysnes.

p. 55 © Arne Nevra.

p. 57 © Casper Tybjerg.

p. 58 © Casper Tybjerg.

p. 58 © Rita Johansen.

p. 59 © Rita Johansen.

p. 60 © Rita Johansen.

p. 62 © Jannike Wika.

p. 63 © Britt Refvik.

SECTION 2   6   p. 67 © Lam Duc Hiên.

p. 70 © Sachiko Haraguchi.

p. 71 © Tara Inniss.

p. 73 © William Cummins.

p. 74 © Frank Fell / The Travel Library / Rex Features / SIPA.

p. 76 © Tara Inniss.

p. 77 © William Cummins.

p. 78 © William Cummins.

p. 79 © William Cummins.

p. 80 © Ministry of Family, Culture, Sports and Youth of Barbados.

7   p. 83 © Ministry of Culture of Morocco.

p. 84 © Ministry of Culture of Morocco.

p. 85 © Editions Gelbart.

p. 86 © Editions Gelbart.

p. 87 © Editions Gelbart.

p. 88 © Editions Gelbart.

p. 89 © Ministry of Culture of Morocco.

p. 90 © Ministry of Culture of Morocco.

p. 91 © Zeppelin / SIPA.

8   p. 95 © Cavalli / SIPA.

p. 96 © SUPERSTOCK / SIPA.

p. 97 © PURESTOCK / SIPA.

p. 98 © Jan Fritz.

p. 99 © Lam Duc Hiên.

p. 161 © Gregory Deagle / Parks Canada.

p. 162 © Brenda Falvey / Parks Canada.

p. 163 © Gregory Deagle / Parks Canada.

p. 165 © SUPERSTOCK / SUPERSTOCK / SIPA.

p. 166 © SUPERSTOCK / SUPERSTOCK / SIPA.

14 p. 170 © Amareswar Galla.

p. 171 © Amareswar Galla.

p. 172 © Amareswar Galla.

p. 173 © Amareswar Galla.

p. 175 © Amareswar Galla.

p. 176 © Amareswar Galla.

p. 177 © Amareswar Galla.

15 p. 179 © Joycelyn B. Mananghaya.

p. 181 © Joycelyn B. Mananghaya.

p. 182 © Joycelyn B. Mananghaya.

p. 183 © Joycelyn B. Mananghaya.

p. 184 © Joycelyn B. Mananghaya.

p. 185 © SA House-Willi / SUPERSTOCK / SIPA.

p. 186 © Stuart Westmorland / SUPERSTOCK / SIPA.

p. 187 © Joycelyn B. Mananghaya.

16 p. 190 © Lassana Cissé.

p. 192 Eliot Elisofon (1970), National Museum of African Art.

p. 193 © Laurent Monlaü.

p. 194 © Our Place World Heritage Collection.

p. 195 © Our Place World Heritage Collection.

p. 196 © Our Place World Heritage Collection.

p. 197 © Laurent Monlaü.

SECTION 4 **17** p. 201 © Rider / SIPA.

p. 205 © Gille de Vlieg.

p. 206 © Our Place World Heritage collection.

p. 208 © Our Place World Heritage collection.

p. 209 © Our Place World Heritage collection.

p. 210 © Our Place World Heritage collection.

p. 211 © Our Place World Heritage collection.

p. 212 © iSimangaliso Wetland Park Authority.

p. 213 © Debbie Cooper / iSimangaliso Wetland Park Authority.

357

p. 269 © Project Management Unit / UNDP Egypt.

p. 270 © Project Management Unit / UNDP Egypt.

p. 271 © Project Management Unit / UNDP Egypt.

p. 272 © Our Place World Heritage Collection.

p. 273 © Our Place World Heritage Collection.

p. 275 © Costanza De Simone.

p. 277 © Spanish Agency for International Development Cooperation.

23   p. 281 © Jan Huneman.

p. 282 © CWSS.

p. 283 © Nordse GmbH / Beate Ullrich.

p. 284 © LKN-SH / Martin Stock.

p. 285 © LKN-SH / Martin Stock.

p. 286 © LKN-SH / Martin Stock.

p. 287 © CWSS.

p. 288 © Focke Strangmann / AP / SIPA.

24   p. 291 © Torfaen County Borough Council.

p. 292 © Torfaen County Borough Council.

p. 294 © Torfaen County Borough Council.

p. 295 © Torfaen County Borough Council.

p. 296 © Phil Rees / Rex Features / SIPA.

p. 299 © Torfaen County Borough Council.

p. 300 © Torfaen County Borough Council.

25   p. 302 © André Pessoa.

p. 303 © Fundação Museu do Homem Americano.

p. 305 © Fundação Museu do Homem Americano.

p. 306 © André Pessoa.

p. 307 © André Pessoa.

p. 308 © Fundação Museu do Homem Americano.

p. 309 © Fundação Museu do Homem Americano.

p. 310 © Fundação Museu do Homem Americano.

p. 311 © Fundação Museu do Homem Americano.

26   p. 313 © APSARA Authority.

p. 314 © Khun Neay Khuon.

p. 315 © APSARA Authority.

p. 316 © Martin Gray.

p. 318 © Our Place World Heritage Collection.

Index

361

365